FOCUS ON MIDDLE SCHOOL CHEMISTRY

Grades 5-8

3rd Edition

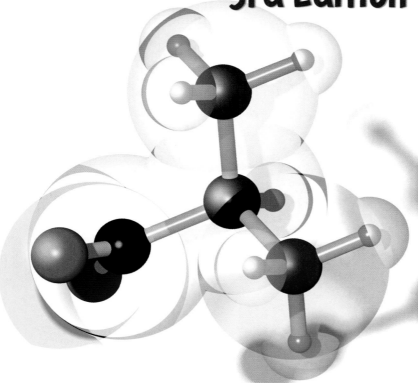

Rebecca W. Keller, PhD

Real Science-4-Kids

Cover design: David Keller
Opening page: David Keller
Illustrations: Rebecca W. Keller, PhD, Janet Moneymaker

Copyright © 2019 Gravitas Publications Inc.

All rights reserved. No part of this publication may be reproduced, stored in a retrieval system, or transmitted, in any form or by any means, electronic, mechanical, photocopying, recording, or otherwise, without prior written permission from the publisher. No part of this book may be reproduced in any manner whatsoever without written permission.

Focus On Middle School Chemistry Student Textbook—3rd Edition (hardcover)
ISBN 978-1-941181-70-6

Published by Gravitas Publications Inc.
www.gravitaspublications.com
www.realscience4kids.com

Contents

CHAPTER 1 WHAT IS CHEMISTRY?   1
  1.1 Introduction   2
  1.2 History of Matter   3
  1.3 The Alchemists   7
  1.4 Alchemy Meets Experiment   8
  1.5 Summary   10
  1.6 Some Things to Think About   11

CHAPTER 2 TECHNOLOGY IN CHEMISTRY   12
  2.1 Introduction   13
  2.2 The Typical Chemistry Laboratory 13
  2.3 Types of Glassware and Plasticware 15
  2.4 Types of Balances and Scales   18
  2.5 Types of Instruments   20
  2.6 Summary   22
  2.7 Some Things to Think About   22

CHAPTER 3 MATTER   23
  3.1 The Atom Today   24
  3.2 The Periodic Table of Elements 25
  3.3 Using the Periodic Table   26
  3.4 Summary   29
  3.5 Some Things to Think About   30

CHAPTER 4 CHEMICAL BONDING   31
  4.1 Introduction   32
  4.2 the Role of Models in Chemistry 32
  4.3 Models of the Atom   33
  4.4 Models of the Chemical Bond   36
  4.5 Types of Bonds   37
  4.6 Shared Electron Bonds   38
  4.7 Unshared Electron Bonds   39
  4.8 Bonding Rules   40
  4.9 Summary   41
  4.10 Some Things to Think About   41

CHAPTER 5 CHEMICAL REACTIONS — 42

- 5.1 Introduction — 43
- 5.2 Chemical Reactions and the Atomic Theory — 43
- 5.3 Types of Chemical Reactions — 45
- 5.4 Combination Reaction — 46
- 5.5 Decomposition Reaction — 46
- 5.6 Displacement Reaction — 47
- 5.7 Exchange Reaction — 48
- 5.8 Spontaneous or Not? — 48
- 5.9 Evidences of Chemical Reactions — 49
- 5.10 Summary — 49
- 5.11 Some Things to Think About — 50

CHAPTER 6 ACIDS, BASES, AND pH — 51

- 6.1 Introduction — 52
- 6.2 Properties of Acids and Bases — 53
- 6.3 Acid-Base Theory — 53
- 6.4 Distinguishing Acids from Bases — 54
- 6.5 Acid-Base Indicators — 57
- 6.6 pH Meters — 58
- 6.7 Summary — 60
- 6.8 Some Things to Think About — 60

CHAPTER 7 ACID-BASE NEUTRALIZATION — 61

- 7.1 Introduction — 62
- 7.2 Titration — 63
- 7.3 Plotting Data — 64
- 7.4 Plot of an Acid-Base Titration — 66
- 7.5 Summary — 70
- 7.6 Some Things to Think About — 71

CHAPTER 8 NUTRITIONAL CHEMISTRY — 74

- 8.1 Introduction — 75
- 8.2 Minerals — 76
- 8.3 Vitamins — 78
- 8.4 Carbohydrates — 79
- 8.5 Starches — 80
- 8.6 Cellulose — 82
- 8.7 Summary — 83
- 8.8 Some Things to Think About — 83

CHAPTER 9 PURE SUBSTANCES AND MIXTURES 84

- 9.1 Introduction 85
- 9.2 Pure Substances—Elements and Compounds 85
- 9.3 What Is a Mixture? 87
- 9.4 Types of Mixtures 88
- 9.5 Solubility of Solutions 91
- 9.6 Surfactants 94
- 9.7 Principles of Separation 96
- 9.8 Techniques of Separation 100
- 9.9 Summary 105
- 9.10 Some Things to Think About 105

CHAPTER 10 ORGANIC CHEMISTRY: THE CHEMISTRY OF CARBON 107

- 10.1 Introduction 108
- 10.2 Hydrocarbons: Alkanes, Alkenes, Alkynes, and Aromatics 111
- 10.3 Alcohols, Amines, Aldehydes, Acids, and Ketones 114
- 10.4 Carbohydrates, Lipids, Fats, and Steroids 117
- 10.5 Summary 122
- 10.6 Some Things to Think About 122

CHAPTER 11 POLYMERS 123

- 11.1 Introduction 124
- 11.2 Polymer Structure 124
- 11.3 Polymer Properties and Reactions 127
- 11.4 Summary 134
- 11.5 Some Things to Think About 134

CHAPTER 12 BIOLOGICAL POLYMERS 135

- 12.1 Introduction 136
- 12.2 Amino Acid Polymers: Proteins 140
- 12.3 Nucleic Acid Polymers 148
- 12.4 RNA 152
- 12.5 Building Biological Machines 153
- 12.6 Summary 156
- 12.7 Some Things to Think About 158

APPENDIX 159

GLOSSARY-INDEX 160

Chapter 1 What Is Chemistry?

1.1 Introduction   2

1.2 History of Matter   3

1.3 The Alchemists   7

1.4 Alchemy Meets Experiment   8

1.5 Summary   10

1.6 Some Things
     to Think About   11

1.1 Introduction

Have you ever wondered what all the objects in the world are made of and why they behave the way they do?

What is soap and why is it slippery? What is air? Why do ice cubes float? Why are dates sweet? What are hair and skin made of? Why is a marble hard and a jellyfish soft?

All of these questions and others like them begin the inquiry into the branch of science we call chemistry.

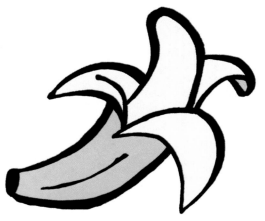

Chemistry is the study of the "stuff" that makes up the things in the physical world. Scientists call this "stuff" matter.

Everything we see with our eyes and can touch with our fingers is made of matter.

Bananas are made of matter. Cars are made of matter. Even our bodies are made of matter. In Chapter 3 we will learn more about matter and what makes up matter.

1.2 History of Matter

The first person we know of who asked questions about matter was Thales, a Greek philosopher who was born in Miletus, a small trading town on the Aegean coast. Thales studied astronomy and mathematics and is believed to have traveled to Egypt where he learned geometry and astronomy. Thales is credited with bringing this knowledge back to Greece. He used what he knew about the stars to his advantage. One story has it that he bought olive presses for making olive oil because he predicted a large olive harvest. He was right! He made lots of money selling olive oil to everyone.

THALES 625-545 BCE

Thales believed that water was the fundamental unit of matter. He thought that everything in the universe came from water. Thales also felt that water could turn into earth and other types of matter.

ANAXIMANDER
611-547 BCE

Another Greek philosopher from Miletus who asked about what things are made of was Anaximander. Many of the philosophers who lived during the time that Anaximander was alive were looking for the essence (the true nature of a thing) of all things. These philosophers were curious about what everything is composed of. Anaximander came up with the idea of "the boundless," or "the ultimate." Unfortunately, he never explained what that was, and this was not a lot of help to people.

ANAXIMENES
Circa 585-525 BCE

Anaximenes was another Greek philosopher who lived in Miletus and wondered what things are made of. In contrast to Thales, Anaximenes believed that air was the basic substance of matter. According to Anaximenes, when air was thinned, it could become fire. In addition, if air was condensed, it would become wind and clouds. And even more condensing would compress air into water, earth, and even stone. Anaximenes tried to explain many natural processes. For example, he believed that thunder and lightning came from wind breaking out of clouds, that rainbows occurred when the Sun's rays hit the clouds, and that earthquakes took place when the ground dried out after a rainstorm.

Empedocles was all things to all people. Some people believed he was a great healer. Others thought he was a magician. He had some convinced he was a living god. Still others believed he was a total fake. The periodic table of earth, air, fire, and water came from Empedocles. He believed that these four "roots" made up all matter. He believed that even living creatures were composed of these materials.

EMPEDOCLES
Circa 490-430 BCE

Another Greek philosopher was Leucippus. We don't know much about Leucippus, but from what we do know, it appears that he was the first person to suggest the idea of empty space. (Today, we would call this a vacuum.) He also developed the idea of atoms. Leucippus believed that different atoms had different sizes and weights. We now know this to be true.

LEUCIPPUS
Circa 480-420 BCE

DEMOCRITUS
Circa 460-370 BCE

Democritus was another Greek philosopher, and he probably was one of the first weather forecasters. Democritus had people convinced that he could predict the future. He was a student of Leucippus, and he is an example of a pupil who is better known than his teacher. He studied many natural objects, and he gave public lectures.

The Greek philosophers debated about a lot of things. One of their debates had to do with sand on the beach. They asked the following questions: Can you divide a grain of sand indefinitely? — and — Is there a point at which you can no longer break the grain in half?

Most of the philosophers believed that you could divide the grain of sand continuously, without ever stopping. Democritus, however, believed that there was a point at which the grain of sand could no longer be broken into smaller pieces. He called this smallest piece of matter the atom. Today we know that atoms make all matter.

The early Greek philosophers had many arguments over the course of many centuries. They argued about how the world works, how it is made, and how it came into being.

As we saw earlier, Thales thought that everything was made of water. He believed that water was the "primary substance" of all things. He thought that water could not be divided any further. Today we know that water is made of two hydrogen atoms and one oxygen atom.

Anaximander rejected water as the primary substance. As we saw earlier, he thought that everything was made of something that he called "the boundless." Nobody was really sure what Anaximander meant by "the boundless," and this made it difficult for him in arguments.

Anaximenes didn't agree with either Thales or Anaximander. He rejected both water and "the boundless" as the primary substance. He believed that air was the primary substance.

Empedocles disagreed with everyone. He said that all of the things in the world are made up of not just one substance but of four — earth, air, fire, and water.

Democritus and Leucippus didn't agree with any of the other philosophers either. Democritus and Leucippus thought that the world was made up of atoms. They had trouble explaining exactly what atoms were because they didn't have the technology to find out about them. However, they thought that all matter is made of one type of thing which they called an atom. They thought that atoms could be combined to make larger things.

It turns out that Democritus's and Leucippus's ideas were closer to reality than the other philosophers' ideas were. But Democritus and Leucippus didn't get very many people to agree with them. Atoms were not seriously considered as a possibility until the 17th century, almost 2000 years later!

1.3 The Alchemists

Many scholars agree that the word chemistry comes from the word alchemy. The word alchemy comes either from the Egyptian word *khemia* which means "transmutation of earth" or from the Greek word *khymeia*, which means the "art of alloying metals." Both word origins point to the alchemists as the first to experiment with chemistry.

Some early experimenters of chemistry were the alchemists. Alchemists were not considered to be true chemists because they did not approach their work with a scientific method. But they did play with the properties of matter. They believed that they could turn some matter, like lead, into different matter, like gold. A lot of what they tried was based on magic and didn't work. In fact, they never got any lead to turn into gold. Often they would go to a king and ask for money to use to make lead turn into gold. Of course, this never happened. Very often, the king would get angry and put the alchemists in prison (or worse). Sometimes the alchemists would just leave town with the king's money.

Although the alchemists were never successful at turning lead into gold, they did learn quite a lot about the properties of matter. They found out which substances would burn, which substances had a particular taste or smell, and which substances would cause bubbles if mixed with other substances. Through this process, they collected lots of information about the properties of various elements.

1.4 Alchemy Meets Experiment

The alchemists didn't think that everything was made of air, water, fire, and earth. They thought that everything was made of mercury, sulfur, and salt! But the alchemists weren't right either. By the late 16th and early 17th centuries, modern scientific thinking began to take shape. Philosophy and invention started coming together, and many philosophers began thinking about how to do quality scientific experiments.

One such thinker was Sir Robert Boyle (1627-1691 CE), an Irish chemist and philosopher. Boyle believed in running experiments to see what would actually happen and to prove or disprove his ideas. He used elaborate glassware to test the properties of air and fire, and by doing these experiments, he figured out fundamental gas laws that describe how gases behave under different conditions. Boyle's experiments also helped show that different kinds of atoms could combine to form molecules.

While doing experiments with air, Boyle produced oxygen, but he didn't know it! However, his experiments led to the later discovery of oxygen as an element. By doing quality experiments, Boyle made many contributions to chemistry.

Joseph Priestley (1733-1804 CE) was an English philosopher and chemist who never took a science course. However, he enjoyed playing around with different materials. After he met Benjamin Franklin, Priestley became very interested in science. He discovered carbon dioxide gas and invented the first soda water by adding carbon dioxide to water, a process called carbonation. Carbon dioxide gas makes the fizz in soft drinks. Another of his many discoveries was nitrous oxide, also called laughing gas, which is used for anesthesia. Priestley is also well known for his experiments with oxygen.

LAVOISIER
1743-1794 CE

Antoine Lavoisier (1743-1794 CE) was a French scientist who believed in performing experiments. He called laboratory work "the torch of observation and experiment." This "torch" shed light on scientific facts. Lavoisier was one of the many scientists who earned the title *The Father of Chemistry*.

Lavoisier knew Priestley, and Priestley told him about his experiments with oxygen (which Priestley called dephlogisticated air). Lavoisier did his own experiments with oxygen and is the one who gave oxygen its name. Lavoisier tried to take credit for the discovery of oxygen, but it was known that others had discovered it before he did.

Lavoisier showed that water is not a basic substance but is made of oxygen and hydrogen. This was a very important discovery for the advancement of chemistry. Lavoisier wrote his ideas and findings in the well known book *Elements of Chemistry,* which contained useful information for chemists of his time and is still available today.

Although Lavoisier's research and discoveries were important to science, he became unpopular during the French Revolution, which was a time of great turmoil. He was taken prisoner and executed.

By the early 1800s, it was well established that air, fire, water, and earth were not the basic substances. This paved the way for the work of John Dalton (1766-1844 CE). Dalton was a British schoolteacher for most of his life, and he first became interested in science by studying the weather.

Dalton revived the hypothesis for the atomic theory of elements that had been proposed by Democritus some 2000 years earlier. In his published work, *A New System of Chemical Philosophy,* consisting of several volumes written between 1808 and 1827, Dalton proposed that all elements are made of atoms. He also proposed that each element has its own atomic weight. The atomic weight, he said, is proportional to the size of the atom that makes up the element. This agrees with what we know today.

Dalton drew the first table of elements. In the table, he described the arrangement of the atoms in several elements, and he provided their atomic weights. Dalton did not know all of the elements that we know today, but he laid the foundation for future study. His contributions to the field of chemistry were significant. John Dalton is known as the *Father of Modern Chemistry*.

Dalton's atomic theory tried to explain some basic properties of atoms. He had the right idea, but several points in his theory were later proven incomplete. Today, we know that atoms make up matter and that the model of the atom is a good explanation for how matter works. Like the alchemists, modern chemists continue to experiment with finding ways to change matter from one type to another. Because they understand about atoms, they've even figured out how to change lead into gold!

1.5 Summary

- Chemistry is the study of the matter that makes up the physical world.
- Early Greek philosophers had different ideas about what matter is made of and had many arguments about how the world works.
- The alchemists experimented with matter and tried to turn lead into gold.
- By doing quality experiments, early scientists were able to show that the basic unit of matter is the atom.

1.6 Some Things to Think About

- Why do you think it is important to ask questions when doing chemistry?

- Describe some of the different ideas philosophers had about matter.

- Why do you think philosophers argued about their ideas?

  Do you think the philosophers' arguments were helpful in the development of theories about matter? Why or why not?

- Even though alchemists are not considered true chemists, how do you think their studies advanced science?

- Explain how the studies of Sir Robert Boyle, Joseph Priestley, Antoine Lavoisier, and John Dalton led to advances in chemistry.

  Do you think each of these scientists could have made the same discoveries without knowing about the work of the others? Why or why not?

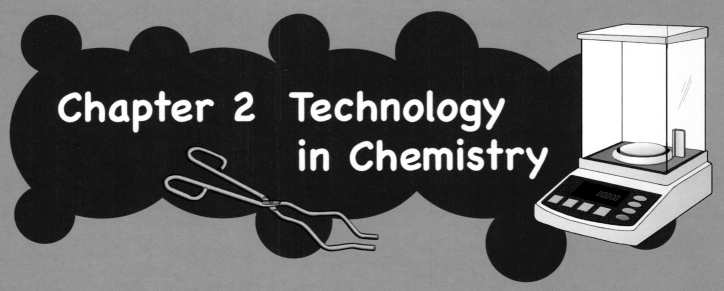

Chapter 2 Technology in Chemistry

2.1 Introduction   13

2.2 The Typical Chemistry Laboratory   13

2.3 Types of Glassware and Plasticware   15

2.4 Types of Balances and Scales   18

2.5 Types of Instruments   20

2.6 Summary   22

2.7 Some Things to Think Abut   22

2.1 Introduction

Chemistry has come a long way from the days of the alchemists and has changed dramatically within the last 100 years. Technological advances have changed how chemists study atoms, molecules, and chemical reactions. Today, chemists can image a single atom, use lasers to discover how molecules move, and work with sophisticated instruments to detect changes in chemical composition.

Lasers being used for analysis
Courtesy of Jim Yost/National Renewable Energy Laboratory (NREL)

The tools used by modern chemists vary depending on the type of chemistry that is being studied. A physical chemist may be interested in the motion of atoms and molecules and may have very different equipment than the organic chemist who is trying to create new compounds. The biochemist may use instruments similar to those found in a biology lab, and an analytical chemist may have a high-end analytical balance not found in a physical chemistry lab. However, whether the lab focuses on biochemistry, organic chemistry, or physical chemistry, most chemistry labs have some common basic equipment.

2.2 The Typical Chemistry Laboratory

Most chemistry labs contain certain basic equipment including different types of glassware for measuring and working with liquids, a balance or scale for measuring solids, a storage rack for chemicals, a stir plate for mixing liquids, bunsen burners with gas outlets for heating liquids and solids, a fume hood for containing sensitive or toxic materials, and safety equipment such as goggles, lab coats, and a safety shower where spilled chemicals can be washed off the body. Most chemistry labs also have long benches made of a material such as epoxy resin that can withstand heavy use and most chemical spills. Alongside the benches are often small desks where research assistants work and keep their notebooks and important papers.

A chemistry lab
Courtesy of Dennis Schroeder/National Renewable Energy Laboratory (NREL)

Some smaller equipment found in most chemistry labs includes various size spatulas and spoons for measuring chemicals, crucibles and crucible tongs for evaporating liquids, funnels, forceps, ring clamps, tube holders, and wash bottles.

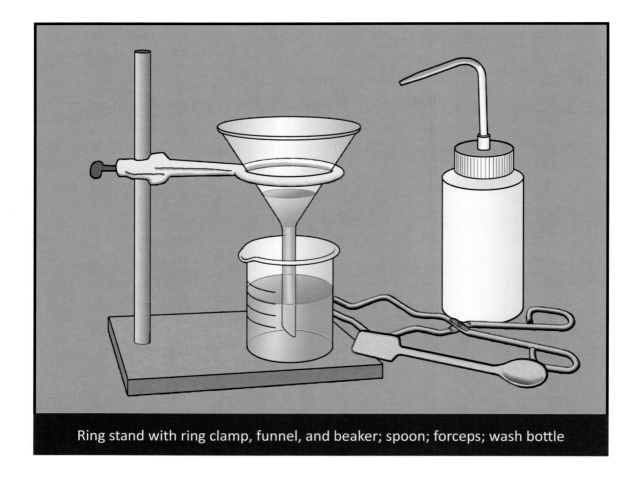

Ring stand with ring clamp, funnel, and beaker; spoon; forceps; wash bottle

2.3 Types of Glassware and Plasticware

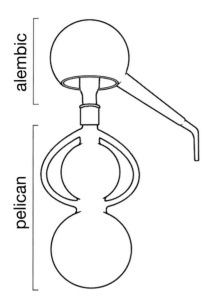

In the days of the alchemists, chemistry experiments were performed in specialized glassware. Two of the first types of glassware used by early chemists were the pelican and the alembic. The pelican and alembic were used for simple distillation, which is a process that separates a liquid mixture into its different individual components.

Distillation occurs when a mixture of liquids is heated and starts to vaporize, turning into a gas. As the vapors rise, they separate, with the lighter weight vapors moving through the long arm and the heavier vapors condensing and pouring back through the side arms.

Many labs now use a modernized version of the pelican and alembic. This version consists of a round-bottomed flask that holds the liquid sample and is attached to a holder, a thermometer, and one end of a condenser. At the other end of the condenser is a collection flask. Since different liquids vaporize at different temperatures, this setup allows chemists to separate liquids by heating the liquid in the sample flask to the temperature at which one of the liquids in the mixture will vaporize. The vapor is returned to the liquid state in the condenser and flows into the collection flask.

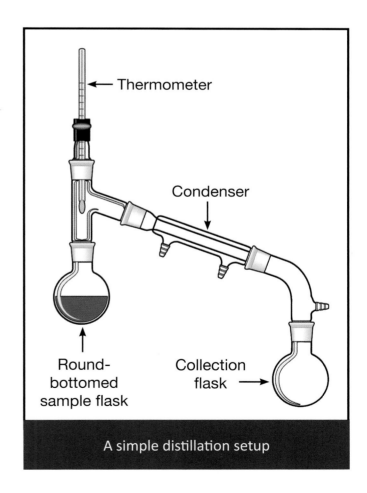

A simple distillation setup

Using beakers in an experiment
Courtesy of Pacific Northwest National Laboratory (PNNL)/DOE

Another type of modern glassware is the beaker. Beakers have a wide mouth and flat bottom and come in a variety of sizes. Beakers are used to measure and transfer large and small amounts of liquids. Beakers typically have markings along the side to indicate different volumes, and they have a spout for pouring.

Flasks are another type of glassware found in many chemistry labs. Flasks have a wide bottom and a narrow mouth. There are two main types of flat bottomed flasks. One type is the Erlenmeyer flask which is cone shaped with a broad, flat bottom and a narrow neck. Erlenmeyer flasks are named after the German chemist Emil Erlenmeyer. The narrow mouth of the Erlenmeyer flask helps minimize spills while mixing and can also be sealed with a rubber stopper.

Erlenmeyer flasks, volumetric flasks, graduated cylinder
Photo courtesy of Warren Gretz/National Renewable Energy Laboratory (NREL)

Chapter 2: Technology in Chemistry **17**

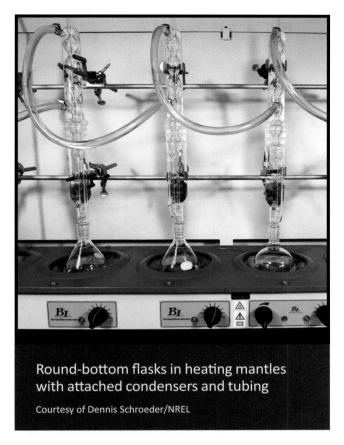

Round-bottom flasks in heating mantles with attached condensers and tubing

Courtesy of Dennis Schroeder/NREL

Another type of flask is the volumetric flask which is round with a flat bottom and tall neck. Volumetric flasks are used to measure a single specific volume. On the neck there is a thin line or ring indicating when the flask is filled to an exact amount.

Round-bottomed flasks don't have a flat area on the bottom. These flasks are used for heating and collecting liquids in a distillation apparatus and can also be attached to other glassware. The rounded bottom helps heat to be equally distributed, especially if the flask is placed in a curved heating mantle.

Graduated cylinders are found in most chemistry labs and are used to accurately measure the volumes of liquids. A graduated cylinder is a tall cylinder made of glass or plastic with markings drawn on the outside to indicate different volumes.

When small volumes or exact volumes need to be measured, a thin tube called a pipet can be used. There are different types of pipets for different purposes. The volumetric pipet is one type. A volumetric pipet is generally a long, thin tube with a bulb in the center. Volumetric pipets are used to measure a single specific volume and are very accurate.

A measuring pipet is a long, thin tube with markings along the side. Measuring pipets deliver various volumes with differing degrees of accuracy.

Transfer pipets are often disposable and generally made of plastic. They are used to transfer small amounts of liquid from one container to another and are not meant to be used for exact measurements.

Very small amounts of liquids can be measured using a pipetman with a plastic tip. A pipetman is a handheld piece of equipment that has a little pump inside that will measure very small amounts of liquids. When the pipetman is held in the palm of the hand and the pumping mechanism is depressed and then released, a small amount of liquid is drawn into the plastic tip. The tip can then be placed in a tube or other container and the liquid pushed out by the pumping mechanism.

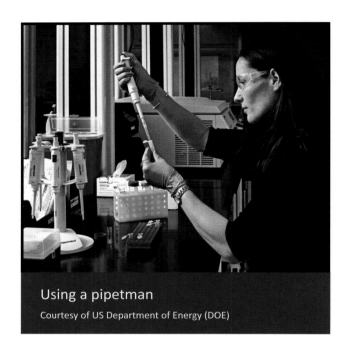

Using a pipetman
Courtesy of US Department of Energy (DOE)

2.4 Types of Balances and Scales

Because chemical reactions require exact measurements of solid, powdered, and liquid materials, one of the most important tools in a chemistry lab is the balance or scale. Most labs use different types of balances or scales depending on the quantity to be measured and the accuracy needed.

The terms *balance* and *scale* are often used interchangeably in chemistry labs, but they are actually slightly different instruments. A balance measures the mass of an object by comparing the amount of force a sample exerts on a lever to the amount of force exerted by a standard reference, or known mass. A scale measures the weight of an object, with weight being related to the amount of the force of gravity that is pulling on the object.

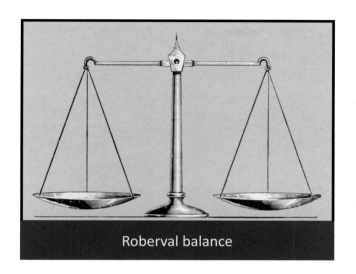
Roberval balance

Because balances measure mass directly, they are considered more accurate than scales. Many research and teaching laboratories have double-pan balances. This type of balance is called a Roberval balance because it was invented by the French mathematician Gilles Personne de Roberval (1602-1675 CE). It has two platforms, one on either side of a pivot point. When one side is heavier than the other, the pan on the heavier side will dip lower than the pan on the lighter side. This type of balance can be used to measure the mass of a sample by adding known reference weights to one pan until the reference weights balance with the sample on the other pan.

Although a balance is considered more accurate than a scale, scales can be used for most applications. Many modern chemistry labs have digital scales. Digital scales give measurements quickly and have an easy to read display. Digital scales can be battery powered or powered with an electric cord.

One type of digital scale used in many labs is the top-loading or pan scale. A pan scale is easy to load and can measure relatively large amounts of both dry and liquid materials. Typically, a measuring boat or beaker is placed on the scale and the scale is set to zero, or "zeroed," with the push of a button. This effectively subtracts the weight of the measuring boat or beaker from the total. Materials can then be added directly to the measuring boat or beaker and the digital pan scale will display the measurement.

If a more accurate measurement is needed and if the sample is small enough, an analytical balance can be used. An analytical balance has a pan mounted on an electromagnetic sensor, both of which are housed in a glass chamber. The glass chamber keeps small air currents from disturbing the measurement.

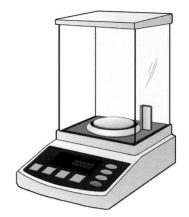

2.5 Types of Instruments

Depending on the type of research being done, many chemistry labs also have a number of specialized instruments that are used to analyze a variety of both chemical and physical properties.

For example, some labs have a gas chromatograph. A gas chromatograph is an instrument that can analyze mixtures of molecules that are volatile and can be turned into a gas. A gas chromatograph has a port where a small sample is injected with a syringe. The sample is heated quickly and turned into a gas. The gas travels down a column where the mixture is separated, then goes through a detector that identifies the gases in the sample. A recorder or computer creates an output showing the different compounds that are in the mixture.

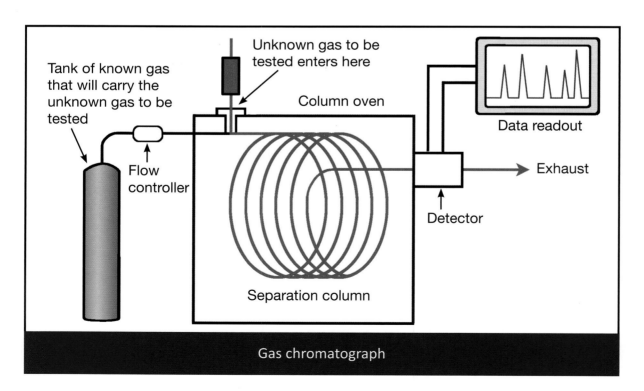

Gas chromatograph

Chapter 2: Technology in Chemistry 21

Using a mass spectrometer
Courtesy of Dennis Schroeder/NREL

The detector in many gas chromatographs is another type of instrument called a mass spectrometer. A mass spectrometer is used to determine the mass of a given sample. In a combined gas chromatograph-mass spectrometer, once the gas has been separated, it enters the mass spectrometer where the gas molecules are bombarded with electrons. Different size atoms will give off different signals when hit with electrons, and these signals can be analyzed. A gas chromatograph-mass spectrometer can be used to test for drugs, toxins in water, and unknown samples.

Another type of instrument used to identify molecules is the infrared spectrometer. An infrared spectrometer measures the vibrations of atoms on a molecule. Molecules with lots of strong bonds and light atoms will vibrate differently than molecules with weak bonds and heavy atoms. This difference can be detected by the infrared spectrometer and can be used to identify different molecules and different types of bonds. Infrared spectrometers are common instruments in labs that study organic molecules.

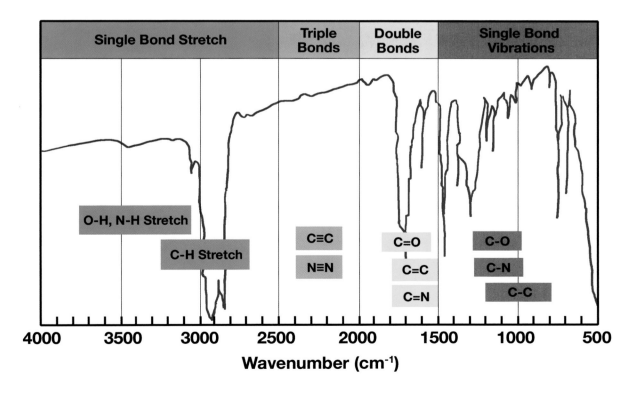

2.6 Summary

- Technology has helped equip modern chemistry labs with different types of glassware, balances, and other instruments.

- Flasks, beakers, and pipets are common types of glassware found in many chemistry labs.

- Double-pan balances, top-loading scales, and analytical scales are used in many chemistry labs.

- Although the type of chemistry research being done determines the instruments that will be needed, many labs have gas chromatographs, mass spectrometers, and infrared spectrometers.

2.7 Some Things to Think About

- What are some tools and equipment you have used when doing chemistry experiments?

- If you had a chemistry lab, do you think many of the same tools and instruments might be used in a geology lab? Why or why not?

- Why do you think flasks come in so many different shapes and sizes?

- If you had a chemistry lab, would you want to have lots of different pipets or would just a few do? Why?

- Why do you think balances and scales are important instruments to have in a chemistry lab?

- When do you think a chemist might want to use a gas chromatograph? What do you think the results would reveal?

Chapter 3 Matter

3.1 The Atom Today — 24

3.2 The Periodic Table of Elements — 25

3.3 Using the Periodic Table — 26

3.4 Summary — 29

3.5 Some Things to Think About — 30

3.1 The Atom Today

Today we know that the fundamental building blocks of matter are atoms. The word atom comes from the Greek word *atomos*, which means "uncuttable."

Today we know that atoms are not really uncuttable but are made of even smaller particles called protons, neutrons, and electrons. However, during chemical reactions atoms act as whole units, so the model of the atom as an uncuttable unit of matter works well for understanding chemical reactions.

Protons and neutrons are roughly equal in size, but electrons are much smaller than either protons or neutrons. Together, protons, neutrons, and electrons make up an atom. Protons and neutrons combine to form the central core (the nucleus) of an atom, and the electrons occupy the space surrounding the central core.

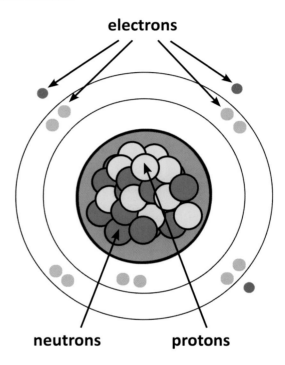

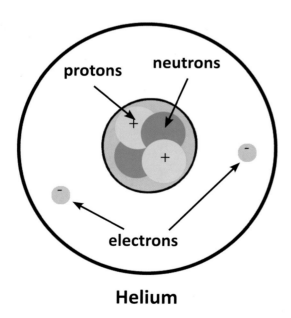

Helium

Electrons have a negative electric charge, protons have a positive charge, and neutrons have no charge. For neutral elements (those that do not have an electric charge) the number of electrons in an atom always equals the number of protons.

Notice that in the helium atom there are two protons and two electrons. Sometimes the number of protons also equals the number of neutrons, as in helium, but this is not always true.

Most of the size of an atom is actually made of the space between the electrons and the core. The protons and neutrons make up only a very small part of the total size.

On the other hand, almost all of the weight of an atom comes from the protons and neutrons. The electrons weigh almost nothing compared to the nucleus.

3.2 The Periodic Table of Elements

The periodic table of elements is a chart used by chemists that categorizes the elements (atoms) and shows their characteristics. Recall from Chapter 1 that the first periodic table of elements was put together by John Dalton (1766-1844) who proposed that all matter is made of atoms.

In 1869 Dmitri Mendeleev, who was a chemist born in Tobolsk, Siberia (Russia), expanded Dalton's table. While scribbling in his notebook, Mendeleev developed the first version of our modern periodic table of elements.

Mendeleev carried with him cards that had the names and weights of the 63 known elements written on them. He thought about the elements and their weights a great deal. After much thinking, he decided to arrange the elements into a chart that was based on their atomic masses.

In 1869 Mendeleev published his chart in a book called *Principles of Chemistry*. He left spaces in his chart because he thought that some elements were missing, and he was right! With his table, he was able to predict a few of the elements that were missing, and while he was still living, the next three elements were indeed discovered. His table gave other scientists the information they needed to find the missing elements. Those missing elements were exactly what Mendeleev predicted! He was famous for the success of his predictions.

Today, the periodic table of elements is much larger than Mendeleev's table and contains 118 elements. Some of the newer elements have been created in the laboratory. The International Union of Pure and Applied Chemistry (IUPAC) is an organization that reviews and verifies the discovery of new elements. As of this writing, the elements ununtrium, ununpentium ununseptium, and ununoctium have been given temporary names until they have been verified and given their official names.

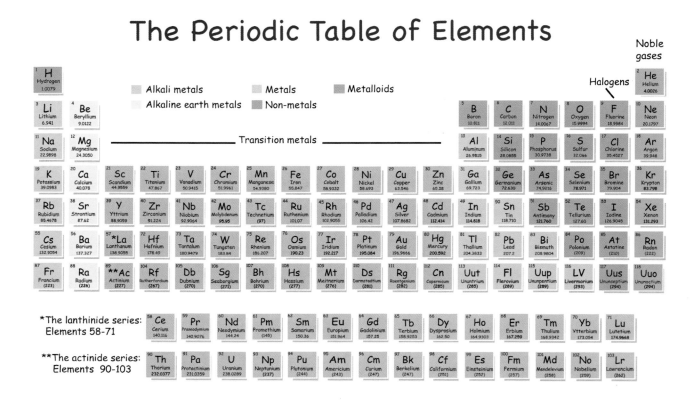

3.3 Using the Periodic Table

In the standard periodic table used by most scientists, the elements are arranged horizontally from left to right in order of increasing atomic number. The atomic number is the number of protons in the nucleus of each atom. For example, carbon has an atomic number of 6. This means carbon has 6 protons in its nucleus. Oxygen has an atomic number of 8, which means it has 8 protons in its nucleus.

Each of the elements has its own symbol. For example, hydrogen has the symbol "H," carbon has the symbol "C," and oxygen has the symbol "O." Notice that for these elements the symbol is the same as the first letter of the name. Other elements have the first two letters of their name as their symbol — for example, "He" for helium and "Ne" for neon.

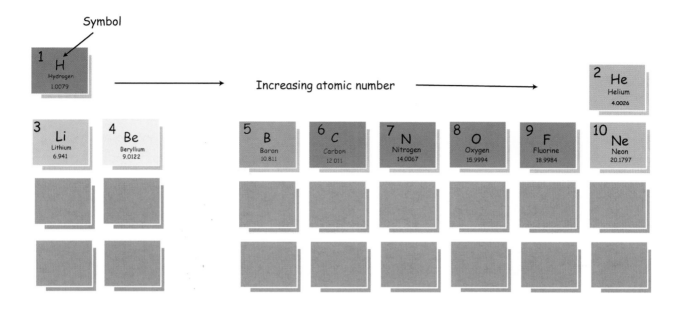

Some elements have a symbol that is different from the first letter or letters of their name. For example, the symbol for gold is "Au" from the word *aurum* which means gold in Latin. The name for sodium comes from the Latin word *natrium*, so sodium has the symbol "Na." Other examples include lead, which has the symbol "Pb" from the Latin word *plumbum*, and tungsten which has the symbol "W" from the German word *wolfram*.

In an atom, the number of protons equals the number of electrons. The atomic number is also the number of electrons in an atom. For example, the smallest element is hydrogen. It has an atomic number of 1, which means it has only one proton. It also has only one electron, since the number of protons equals the number of electrons.

Though atoms are very small, each one has a weight called the atomic weight. For most atoms the atomic weight is very close to the sum of the protons and neutrons in the nucleus. Protons and neutrons each have an atomic weight of 1. Electrons are so small that they are considered to have almost no weight at all. The number of neutrons for an atom can be calculated by subtracting the number of protons from the atomic weight.

For example, the atomic weight of hydrogen is 1.0079, which on this periodic table is the number found below the name. To find the number of neutrons, the number of protons (1) is subtracted from the atomic weight (1.0079 or 1); 1 - 1 = 0. This means that hydrogen has no neutrons and only one proton in its nucleus, or core.

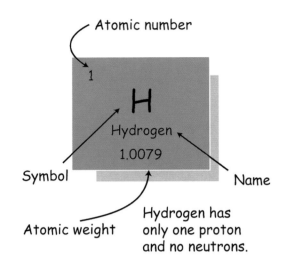

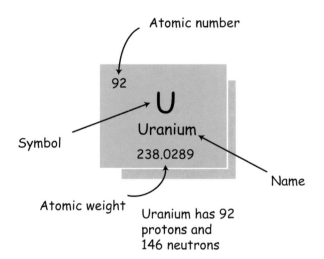

The largest naturally occurring element is uranium. It has an atomic number of 92, which means it has 92 protons and 92 electrons. It has an atomic weight of 238.0289. To calculate the number of neutrons, the number of protons is subtracted from the atomic weight (238 - 92 = 146), so uranium has 146 neutrons.

The elements in the periodic table are arranged vertically according to their chemical properties. All of the elements in a single column undergo similar chemical reactions and have similar chemical properties. All of the elements in the far right-hand column are called the noble gases. They are similar to each other because they don't react with other atoms or molecules. The elements in the far left-hand column are called the alkali metals. They are similar to each other because they react with lots of different atoms and molecules.

Chapter 3: Matter   29

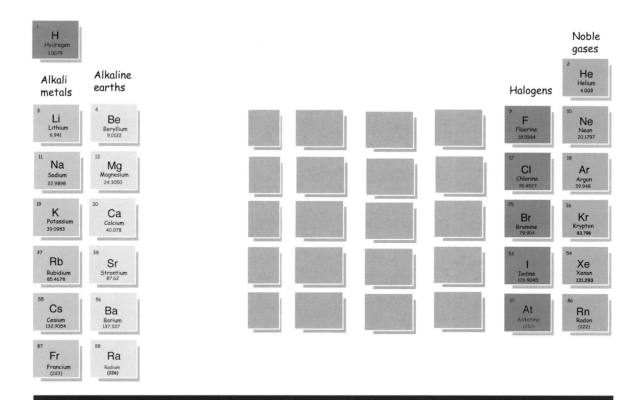

All of the elements in a single column have similar properties.

The periodic table of elements organizes a lot of information about the elements and their chemical properties. This table helps chemists predict the behavior of the elements and how they might interact with each other.

3.4 Summary

- All things, both living and nonliving, are made of atoms, which are also called elements.

- Atoms are made of protons, neutrons, and electrons.

- In an atom, the number of protons equals the number of electrons.

- All atoms (elements) are found in the periodic table of elements.

- The elements (atoms) are arranged in the periodic table in groups that have similar properties.

3.5 Some Things to Think About

○ Why do you think it is helpful to chemists to have an atomic model that represents the structure of a specific atom?

○ Do you think the periodic table is complete, or do you think there may be elements that have not yet been discovered? Why?

○ List the information you can discover about nitrogen by looking at the periodic table.

How do you think these facts could be useful to you in doing chemistry?

Chapter 4 Chemical Bonding

4.1	Introduction	32
4.2	The Role of Models in Chemistry	32
4.3	Models of the Atom	33
4.4	Models of the Chemical Bond	36
4.5	Types of Bonds	37
4.6	Shared Electron Bonds	38
4.7	Unshared Electron Bonds	39
4.8	Bonding Rules	40
4.9	Summary	41
4.10	Some Things to Think About	41

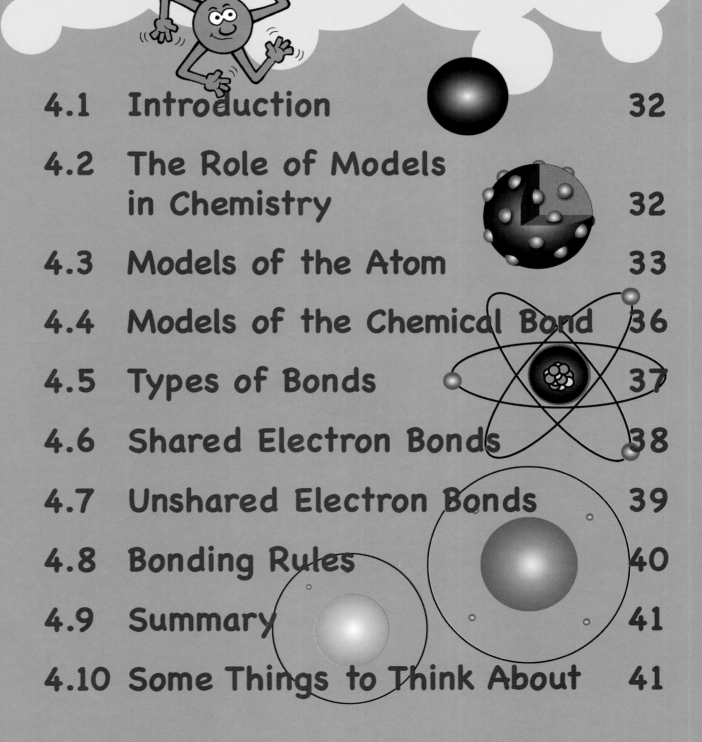

4.1 Introduction

As we saw in the previous chapter, everything is made of atoms. Sometimes substances such as gold and silver are made of a single type of atom. Sometimes substances are made of more than one type of atom. For example, table salt is a combination of sodium and chlorine atoms. When two or more atoms combine with each other, they make a molecule. But how do atoms stick together, or bond, to make a molecule?

4.2 The Role of Models in Chemistry

Before we look at bonds, let's first explore models in chemistry. A model is simply an idea of how something looks or works. Our understanding of a chemical bond is based on models of how electrons interact with each other.

Anything can be represented by a model. Maybe you like to build model airplanes, model ships, or model cars. Even though a model can represent an airplane, ship, or car, we know the model is not the real thing. We understand that models give us an idea of what the real thing looks like, and that idea helps us to understand more about the real thing.

A model in science is the same. We have an idea of what something is, how it is structured, and how it works. We use models to help us think about and further develop our ideas. Models also help us explain our ideas to other people. In science, a model is usually very incomplete and does not have all the details of the real thing. In fact, many times scientists don't know exactly what the real thing looks like!

Even so, building models helps scientists better understand how things work and helps in designing experiments. Running experiments is important in science, and models help scientists decide what experiments to do. We can test the model by doing experiments that will show whether the model fits with the data.

4.3 Models of the Atom

Building models of the atom has been a very important part of chemistry for many years. There have been many different models proposed as new information about atoms has been discovered.

For example, John Dalton was not aware of protons, neutrons, or electrons. He perceived the atom as a small hard sphere, like a billiard ball. This model of the atom is still useful and is used in many illustrations of atoms and molecules. However, because it does not show how electrons are arranged, it is not useful for understanding how atoms form bonds.

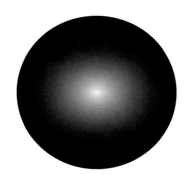

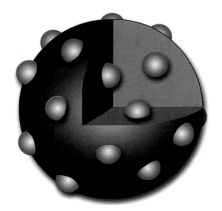

In 1897 Sir Joseph John Thomson (1856-1940 CE) proposed a different model. Thomson was a British physicist who studied electricity and how magnetic fields affect the path of light. Through his research Thomson discovered the electron and its very small size. Thomson proposed a model similar to the billiard ball model but with the electrons randomly embedded. The model looks like plum pudding and is referred to as the plum pudding model.

However, it was later shown that electrons don't sit randomly around the atomic core, so the model needed to be revised again.

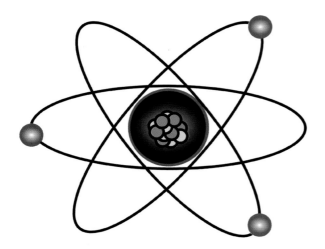

In the early 1900s Ernest Rutherford (1871-1937 CE), a British physicist, found that almost all of the mass of an atom is located in the center with the majority of the atom consisting mainly of empty space. Rutherford's finding contributed to a model proposed by Niels Bohr (1885-1962 CE), a Danish physicist, showing that the electrons circle the atomic core in fixed orbits with the low energy electrons orbiting close to the center and the high energy electrons orbiting farther away.

However, as scientists continued to explore atoms, electrons, protons, and neutrons, the model of the atom continued to grow more complex.

Today, scientists use quantum mechanics to understand how electrons move and form bonds. Quantum mechanics uses both physics and math. Sometimes the research done in quantum mechanics results in very strange ideas about how small particles such as electrons behave. Because it's impossible to know both the position of an electron and how fast it is going at exactly the same time, quantum mechanics tells us that electrons exist in orbits as probabilities! In other words, we can only describe how likely (how probable) it is that an electron is located at a specific position around an atom. This means that electrons don't circle the atom in a fixed orbit as Bohr suggested.

Chapter 4: Chemical Bonding 35

Carbon

In the *Real Science-4-Kids* textbooks, we also use different ways to model the atom. In *Focus On Elementary Chemistry* we modeled the atom as a ball with arms and a face. The arms represent the electrons that are available for bonding and show how they link together to form molecules.

Electrons make the bonds that join atoms together to form molecules. To show this, a new model was needed. The arms in our models were replaced by red dots to represent the electrons that can form bonds, and gray dots were added to represent the electrons that can't form bonds.

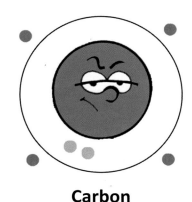
Carbon

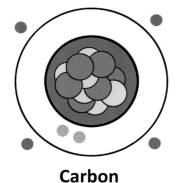
Carbon

Now that we've learned about protons and neutrons, we can add these to our model. We can take away the face and show how protons and neutrons occupy the nucleus, or atomic core.

However, because protons and neutrons don't participate in bonding, we don't need to show them in our model when we are illustrating how atoms make molecules. We can replace the core with a hard sphere or billiard ball, similar to Dalton's model, and keep the bonding electrons. We will use this model to illustrate how atoms form bonds, and for larger molecules and chemical reactions, we will use the hard sphere.

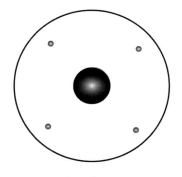
Carbon

All of these different models focus on different features. Some models focus on the number of bonds. Some models focus on the protons and neutrons. Some models focus on both the bonding and non-bonding electrons. None of these models illustrate exactly what an atom looks like, but they can still be used to understand atoms and how atoms form bonds.

4.4 Models of the Chemical Bond

Once scientists knew about electrons in atoms, they could ask how atoms were held together to form molecules. The idea of atoms connecting to form molecules was thought about for centuries. René Descartes (1596-1650 CE), a French philosopher, proposed a model of the chemical bond and believed that molecules were held together by little hooks and eyes. He believed that some atoms had hooks and that others had eyes where the hooks could connect. Today we know that atoms don't have hooks and eyes, but Descartes' model added to our understanding of chemical bonds.

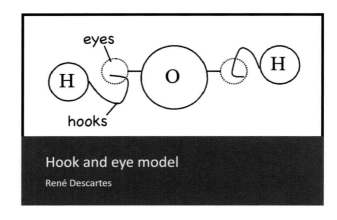

Hook and eye model
René Descartes

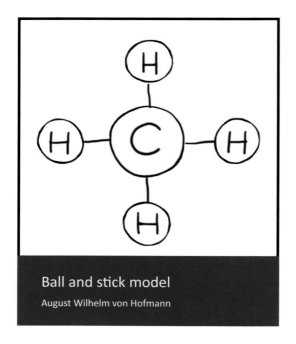

Ball and stick model
August Wilhelm von Hofmann

August Wilhelm von Hofmann (1818-1892 CE), a German chemist, was the first person to build models of molecules instead of just drawing them. He used sticks for chemical bonds and balls for the individual atoms. He did not know the real shapes, but he gave us a start in understanding the science of molecular shape.

The ball and stick model is still in use today, but it does not give a complete idea about a molecule because the model is flat. We need to see a molecule in three dimensions to really understand how it will react with other molecules and atoms.

Hermann Emil Fischer (1852-1919 CE) was a German chemist who proposed some ideas about molecular shape for enzymes, which are a type of molecule found in living things. He also developed a way of drawing on paper a two-dimensional representation of a three-dimensional molecule. In a Fischer projection all of the bonds are drawn as vertical or horizontal lines. For methane, the carbon atom is represented by the intersection of the horizontal and vertical lines. A Fischer projection is a representation of a three-dimensional molecule. The horizontal lines represent bonds that come forward, out of the page.

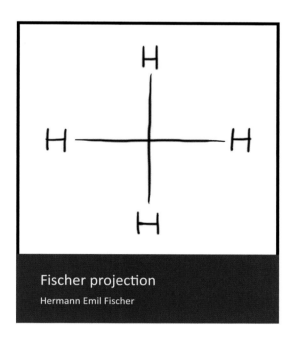

Fischer projection
Hermann Emil Fischer

4.5 Types of Bonds

Today we know that there are two general types of bonds: bonds between atoms where the electrons are shared and bonds between atoms where the electrons are not shared.

Molecules that have shared electron bonds behave very differently from molecules that have unshared electron bonds. This difference is quite important because it determines the way molecules interact with other molecules.

4.6 Shared Electron Bonds

Hydrogen is an example of an atom that has one electron available for bonding. When two hydrogen atoms bond to form a molecule, they form a bond where the electrons are equally shared. Because the atoms are identical (they are both hydrogens), one atom cannot take more electrons for itself than the other atom. This results in a bond with shared electrons, and in this particular case the electrons are equally shared. A bond with shared electrons is a covalent bond.

Bonds that are equally shared are *always* formed between two identical atoms, such as two hydrogens, two oxygens, two nitrogens, or two chlorine atoms.

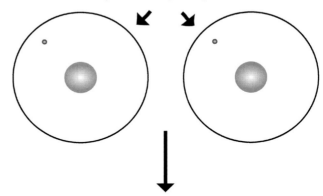

Two separate hydrogen atoms

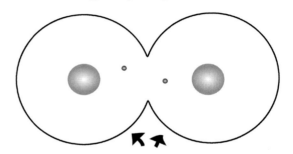

One single hydrogen molecule

Space shared by both electrons

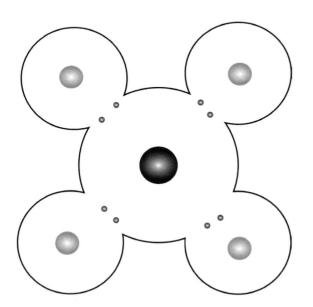

A carbon atom with four hydrogens sharing their electrons (methane)

However, covalent bonds also form between atoms that are not identical but still want to share electrons — such as carbon and oxygen, carbon and hydrogen, and hydrogen and oxygen. The molecule (methane) formed by one carbon atom and four hydrogen atoms also has bonds with shared electrons that are covalent bonds.

4.7 Unshared Electron Bonds

Sometimes the electrons are not shared between two atoms. This happens when there is one atom that wants to get more electrons for itself and is not willing to share its electrons, and the second atom wants to give electrons away. This results in a bond with *unshared* electrons. This type of bond is called an ionic bond. The word *ionic* in this case means that the atoms form ions. An ion is an atom that is positively or negatively charged.

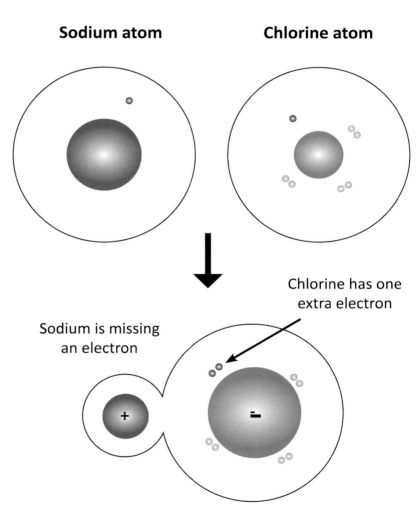

The electrons in the bond between a sodium atom and a chlorine atom are *not* shared. The sodium atom gives away one of its electrons to the chlorine atom. As a result, the sodium atom has fewer electrons than it would normally have if it were alone, and the chlorine atom has more electrons than it would normally have if it were alone.

The molecule formed by sodium and chlorine is called sodium chloride.

4.8 Bonding Rules

The number of electrons an atom has available for bonding also determines how many bonds an atom can form. An atom cannot form just any number of bonds with another atom. For example, a hydrogen atom has only one electron, so it usually forms only one bond. Carbon, on the other hand, has a total of 6 electrons. However, only 4 of those electrons are available to make bonds, so carbon atoms usually form a total of 4 bonds. Nitrogen has only three available electrons and usually forms three bonds. Oxygen, with two available electrons, typically forms only two bonds.

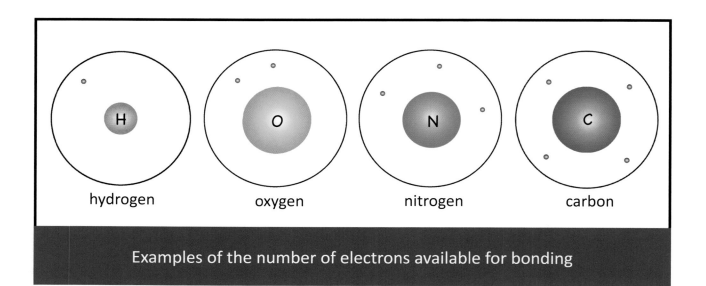

Examples of the number of electrons available for bonding

4.9 Summary

- Electrons join together to form chemical bonds between atoms.
- Building models helps scientists understand how things work and how they are made.
- As scientists learn more about electrons and bonding, models change.
- A covalent bond occurs when atoms share electrons.
- An ionic bond occurs when atoms do *not* share electrons.
- Different atoms have different numbers of electrons available for bonding.

4.10 Some Things to Think About

- What do you think happens as a result of atoms combining to make a molecule?
- Describe a model you have made. Was your model an object, a drawing, or something else? What was complete about it and what was incomplete? How did it help you better understand the object you were modeling?
- Why do you think scientific models change over time?
- How do you think the different types of Real Science-4-Kids models of atoms reflect your expanding knowledge of chemistry?
- Why do you think models of molecules can be helpful to chemists who are studying chemical bonding?
- Do you think chemists always use the same type of model? Why or why not?
- What are the two general types of bonds?
- Do you think it can be helpful to chemists to know how different types of atoms will bond? Why or why not?
- What do you think is needed for atoms to form a covalent bond?
- How are ionic bonds formed? How are they different from covalent bonds?
- Why do you think a chemist would find it important to know how many bonds a certain type of atom can form?

Chapter 5 Chemical Reactions

5.1   Introduction   43

5.2   Chemical Reactions and the
      Atomic Theory   43

5.3   Types of Chemical Reactions   45

5.4   Combination Reaction   46

5.5   Decomposition Reaction   46

5.6   Displacement Reaction   47

5.7   Exchange Reaction   48

5.8   Spontaneous or Not?   48

5.9   Evidences of Chemical Reactions   49

5.10 Summary   49

5.11 Some Things to Think About   50

Chapter 5: Chemical Reactions

5.1 Introduction

What happens inside a battery? How does a car use gasoline to run? What happens to an egg when it's fried in a pan? All of these situations are examples of chemical reactions.

Chemical reactions happen everywhere! When we look at the world around us, we see thousands of chemical reactions happening all the time. We see plants using water and sunlight to grow. We see cars using fuel to move. We see foods being cooked, and factories making fabrics. All of these activities require chemical reactions.

People were using chemical reactions even before history was recorded. For example, ancient people used animal skins to make clothes. These people treated animal skins in order to make them soft. During the treatment of the skins, chemical reactions occurred. They didn't know what the reactions were, but they knew that by treating the skins they could change them.

Ancient farmers who grew crops for food discovered other chemical reactions. They knew that there were materials that they could put in the ground, and they knew that those materials would help the plants grow. Today, farmers use chemical reactions to help plants grow and keep pests away!

5.2 Chemical Reactions and the Atomic Theory

By the time John Dalton came up with his atomic theory, there were already some facts known about chemical reactions. People knew that materials could be changed into other materials, but it looked like substances were lost during those changes. For example, when wood is burned, ashes are left. However, the ashes weigh a lot less than the wood did before it was burned. It looks as if a lot of matter is lost during this chemical reaction.

Early chemists didn't know that matter isn't lost when wood turns to ash. Dalton and others realized that burning wood gives off gases. We can't see these gases, but we can show that they exist. When wood burns, part of the chemical reaction involves making gases, and the gases have some of the atoms that were in the original wood.

Recall that part of Dalton's atomic theory involved the idea of molecules. Dalton said that atoms are neither created nor destroyed in a chemical reaction — the atoms are just rearranged into different combinations.

The understanding that atoms are neither created nor destroyed during the process of a chemical reaction is called the law of conservation of matter. The law of conservation of matter was originally proposed by Antoine Lavoisier, the French chemist. By trapping the oxygen that is released when mercuric oxide (a compound of mercury and oxygen) is heated, Lavoisier proved that matter is not lost during a chemical reaction.

The law of conservation of matter is very important in chemistry. Because we know how many and what type of atoms a chemical reaction starts with, we also know how many and what type of atoms are present when the chemical reaction is complete.

The starting materials in a chemical reaction are called reactants and the ending materials are called products.

Because atoms are neither created nor destroyed, all the atoms we have in the universe today are the same atoms that existed thousands of years ago! It's strange to imagine that the oxygen atoms we are breathing could be the same oxygen atoms breathed by George Washington, Mendeleev, Cleopatra, or Ibn al-Haytham. They could even be the oxygen atoms that Democritus breathed long ago!

Chapter 5: Chemical Reactions   45

It's even more strange to think about where the atoms in our bodies came from. In your body, you could have carbon atoms that were once in the body of Julius Caesar, the great Roman emperor. Or maybe you have atoms in you that were once in an eagle that flew over the mountains thousands of years ago.

5.3 Types of Chemical Reactions

A chemical reaction occurs:

> whenever bonds between atoms and molecules are created or destroyed.

Whenever two atoms, two molecules, or an atom and a molecule interact with each other and cause bonds to be created or destroyed, a chemical reaction has occurred.

Because there are many different kinds of chemical reactions, it is useful to categorize them. There are different ways to categorize chemical reactions, but to make it simple we will look at the following four general types:

1. Combination reaction: Occurs when two or more molecules combine with each other to make a new molecule.

2. Decomposition reaction: Occurs when a molecule decomposes, or breaks apart, into two or more molecules.

3. Displacement reaction: Occurs when one atom kicks another atom out of a molecule.

4. Exchange reaction: Occurs when one atom trades places with another atom in a molecule.

5.4 Combination Reaction

In a combination reaction, two or more molecules combine to form a single product.

The reaction of sodium and chlorine to make sodium chloride (table salt) is an example of a combination reaction. In this reaction two sodium atoms combine with one molecule of chlorine gas to make two molecules of sodium chloride. The sodium and chlorine atoms are the reactants and the sodium chloride is the product.

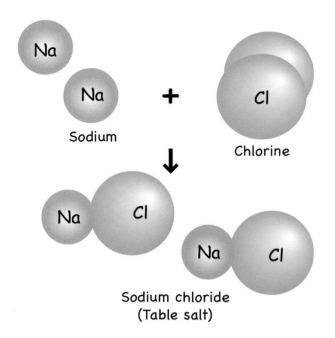

5.5 Decomposition Reaction

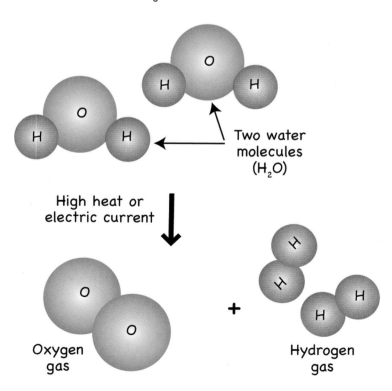

In a decomposition reaction, molecules of one type break apart, or *decompose,* to make two or more products. The breakup of water into hydrogen and oxygen gases is an example of a decomposition reaction.

5.6 Displacement Reaction

A third general type of chemical reaction is the displacement reaction.

In this reaction, one atom will remove another atom from a compound to form a new product.

The formation of sodium hydroxide from two water molecules and two metallic sodium atoms is an example of a displacement reaction. The sodium atoms (labeled "Na" and shown as blue balls) kick out hydrogen atoms (labeled "H" and shown as gray balls) from the water molecules. The sodium atoms combine with the remaining oxygen atom and hydrogen atom from the water molecule (called a hydroxide ion) to make a new molecule called sodium hydroxide. The two hydrogen atoms that were kicked out by the sodium atoms form a hydrogen gas molecule.

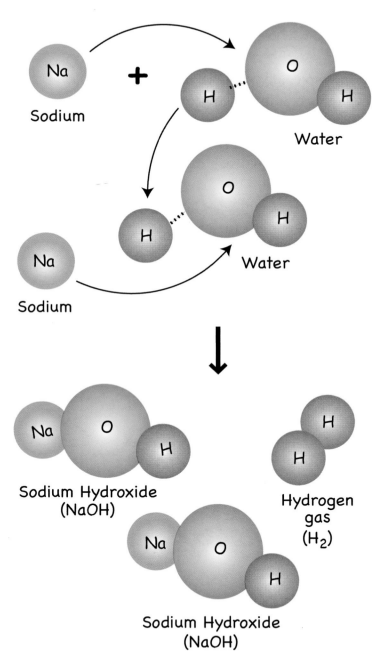

5.7 Exchange Reaction

The fourth type of general chemical reaction is the exchange reaction. In this reaction, the atoms of one molecule trade places with the atoms of another molecule to form two new molecules.

The reaction of hydrochloric acid (HCl) and sodium hydroxide (NaOH) is an example of an exchange reaction. The hydrogen atom in the HCl molecule trades places with the sodium atom in the NaOH molecule to make two new molecules, sodium chloride (NaCl, or table salt) and water (H_2O).

These are the basic types of simple chemical reactions. Some reactions that have many components are much more complicated than those outlined in this chapter, but most chemical reactions fall into one of these four categories.

5.8 Spontaneous or Not?

Not all chemical reactions are spontaneous. Spontaneous means the reaction happens all by itself by just mixing the chemicals. The reaction of hydrochloric acid (HCl) and sodium hydroxide (NaOH) (an exchange reaction) is a spontaneous reaction. However, not all chemical reactions are spontaneous. The decomposition reaction of water into hydrogen gas and oxygen gas is *not* a spontaneous reaction. It requires either high heat or an electric current. It's a good thing that not all reactions occur spontaneously. Imagine how difficult it would be to swim or sail a boat, or even get a drink, if water spontaneously decomposed into hydrogen gas and oxygen gas!

5.9 Evidences of Chemical Reactions

To determine whether or not a chemical reaction has occurred, chemists look for certain signs, or evidences, of a change. There are several signs that tell scientists when a chemical reaction has occurred.

A chemist may look for bubbles being released when something gets added to something else. Bubbles indicate that a gas formed during the reaction.

Or, if two solutions are mixed, one solution might change color.

Sometimes when two solutions are mixed, a temperature change occurs, and the solution gets either hotter or colder.

Finally, another indication that a chemical reaction has taken place is the formation of a precipitate, which can look like colored sand, mud, dust, or snow forming in a solution. A precipitate forms when new molecules being made from the chemical reaction do not dissolve in the solution.

These are just some of the ways that chemists can tell when a chemical reaction has taken place.

Bubbles

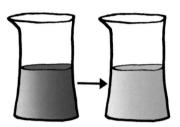
Color change

Temperature change

Precipitation

5.10 Summary

- The law of conservation of matter states that matter is neither created nor destroyed during a chemical reaction.

- A chemical reaction occurs whenever bonds between atoms and molecules are created or destroyed.

- There are different kinds of chemical reactions. Four of these are as follows:

    Combination reactions occur when molecules join.

    Decomposition reactions occur when molecules break apart.

    Displacement reactions occur when molecules are removed.

    Exchange reactions occur when molecules trade places.

- Not all chemical reactions occur spontaneously.

- Sometimes changes occur that indicate a chemical reaction has taken place. These changes include bubble formation, color changes, temperature changes, and precipitates.

5.11 Some Things to Think About

- List some things you do or use every day that require chemical reactions.

- Why do you think the law of conservation of matter is important for chemists?

- If two substances are mixed together and have a chemical reaction, will there always be a chemical reaction when those substances are mixed? Why or why not?

- Do you think sodium chloride is like either sodium or chlorine? Why or why not?

- Explain the difference between a combination reaction and a decomposition reaction.

- In the example illustrated in Section 5.6, are the same atoms present before and after the displacement reaction? How are the atoms arranged in molecules before and after the reaction?

- How would you describe an exchange reaction in your own words? How do you think it is different from the other types of reactions you have learned about?

- Do you think a spontaneous reaction causes batter to tun into a cake? Why or why not?

- What examples of spontaneous reactions can you think of?

- List some examples of evidence of chemical reactions you have observed.

Chapter 6 Acids, Bases, and pH

6.1 Introduction　　　　　　　　　　　　　52

6.2 Properties of Acids and Bases　　　53

6.3 Acid-Base Theory　　　　　　　　　53

6.4 Distinguishing Acids from Bases　54

6.5 Acid-Base Indicators　　　　　　　57

6.6 pH Meters　　　　　　　　　　　　　58

6.7 Summary　　　　　　　　　　　　　60

6.8 Some Things to Think About　　　60

6.1 Introduction

Modern technology has allowed chemists to explore not only the structure, size, and properties of atoms and molecules but also how atoms and molecules interact with each other during chemical reactions. Two specific types of molecules that play a role in many important chemical reactions are acids and bases.

Even before people knew what acids and bases were, these compounds were used for many different purposes. Tablets from the Babylonian culture show that people knew how to make soap from bases 2800 years ago. Some of the Babylonian people and ancient German tribes used soap to style their hair. The base they used for soap making is called lye, which is a substance found in the ashes left over after wood has burned. Different kinds of oils or animal fats were then heated with the lye, producing soap.

ABU MUSA JABIR IBN HAYYAN
Circa 721-815 CE

Some acids were also known many centuries ago. An Iranian alchemist named Abu Musa Jabir ibn Hayyan (circa 721-815 CE) discovered hydrochloric acid by mixing a salt (sodium chloride) with sulfuric acid. Jabir ibn Hayyan also developed aqua regia [Latin, royal water] by mixing nitric acid and hydrochloric acid. This material could easily dissolve gold and was often used to determine whether or not a substance appearing to be gold was, in fact, gold.

Chapter 6: Acids, Bases, and pH   53

6.2 Properties of Acids and Bases

Some general properties of acids and bases are listed in the following chart. Vinegar, tomatoes, and black coffee are all acids, and all have a sour taste. It turns out that most acids are sour tasting. Grapefruit, for example, can be very sour. The juice inside a grapefruit contains a lot of citric acid.

General Properties	
Acids	Bases
• Sour in taste • Not slippery to the touch • Dissolve metals	• Bitter in taste • Slippery to the touch • React with metals to form precipitates

Detergents and many cleaners feel slippery. This is because many cleaners are basic and slipperiness tends to be a property of bases.

Some acids and bases are very poisonous or corrosive and can easily harm you. Battery acid, for example, can burn your skin if it happens to spill and can also make you very sick if eaten.

6.3 Acid-Base Theory

Because acids and bases are so important, chemists have developed several ways of understanding them. In 1834, Michael Faraday (1791-1867 CE) discovered that acids and bases are electrolytes, meaning they form ions when dissolved in water and can conduct electricity. Ions are atoms or molecules that have an electric charge.

Svante Arrhenius (1859-1927 CE), a Swedish chemist, took the next step in understanding acids and bases. In 1884 Arrhenius showed that acids produce hydrogen ions (H+) in water and bases produce hydroxide ions (OH-) in water. This is a useful theory for explaining acids and bases, and

SVANTE ARRHENIUS
1859-1927 CE

the Arrhenius definitions are still widely used today. By definition, an Arrhenius acid is any molecule that releases a hydrogen ion (H+), and an Arrhenius base is any molecule that releases a hydroxide ion (OH-). When using this definition, keep in mind that it only applies to hydrogen and hydroxide ions in aqueous (water) solutions. Acids and bases can also be defined as the giving or taking of protons or electrons in non-aqueous (non-water) solutions. However, for the chemical reactions explored in this textbook, the Arrhenius definition of an acid and base will be used.

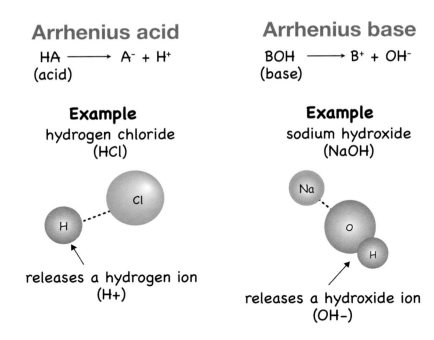

6.4 Distinguishing Acids from Bases

The use of litmus paper was the first method discovered for determining whether a liquid is an acid or a base. The word litmus comes from an old Norse word meaning "to dye or color."

Certain species of lichens provide the dye used in litmus paper. A lichen is an organism that consists of a fungus and algae working in partnership to form the organism. Litmus paper was first used by the Spanish alchemist Arnaldus de Villa Nova (circa 1235-1311 CE) to test whether a substance was an acid or a base.

Blue litmus paper will turn red in the presence of an acid and red litmus paper will turn blue in the presence of a base. Because litmus paper is relatively inexpensive to produce and easy to use, it can be used in the chemistry lab to quickly determine whether an aqueous solution is an acid

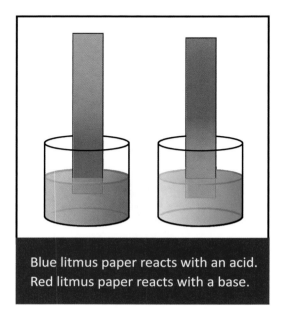

Blue litmus paper reacts with an acid. Red litmus paper reacts with a base.

or a base. Litmus paper is a great tool for chemical geologists when they are out in the field and need an inexpensive and lightweight method for testing the water in ponds, rivers, and geothermal pools.

Litmus paper can be perfect for determining whether a solution is an acid or a base, but it cannot measure how concentrated an acid or a base is. Concentration is defined as the number of units in a given volume. The definition of a unit can vary, and the unit can be an atom, an electron, or even a ping-pong ball. A solution that contains many units is called concentrated (or strong), and a solution with few units is called dilute (or weak). For example, a concentrated solution of hydrochloric acid (HCl) has many HCl molecules, and a dilute solution of HCl has few HCl molecules.

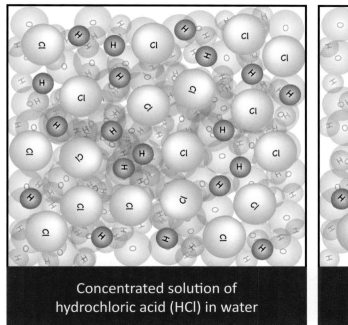
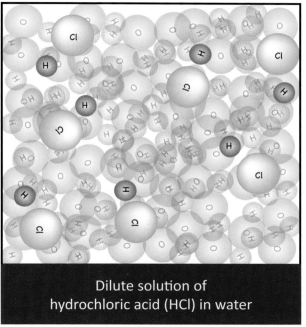

Concentrated solution of hydrochloric acid (HCl) in water

Dilute solution of hydrochloric acid (HCl) in water

The concentration of an acid or base depends on the number of acid or base ions in a solution. Using the Arrhenius definition, a concentrated acid is an acid that contains a large number of hydrogen ions (H+) and a concentrated base is a base with a large number of hydroxide ions (OH-). Conversely, a dilute acid is an acid with few hydrogen ions and a dilute base is a base with few hydroxide ions. But how do you measure the number of hydrogen or hydroxide ions in a given volume?

In 1909 while working for a brewery in Sweden, Sören Peter Lauritz Sörensen (1868-1939 CE), a Danish chemist, introduced the pH scale (pH is pronounced "P" "H"). The pH scale makes it easier for chemists to describe how many hydrogen ions are in a solution, and therefore, how acidic or basic a solution is. For the brewery, knowing the pH of their mixtures enabled them to control the acidity so they could make a better and more consistent product.

Specifically, pH is a measure of the concentration of hydrogen ions in a solution. According to the pH scale, an acid has a pH below 7 and a base has a pH above 7. Neutral water has a pH equal to 7.

pH = 7: The solution is neither an acid nor a base — it is neutral.

pH less than 7: The solution is an acid.

pH more than 7: The solution is a base.

The following chart shows the pH for different solutions.

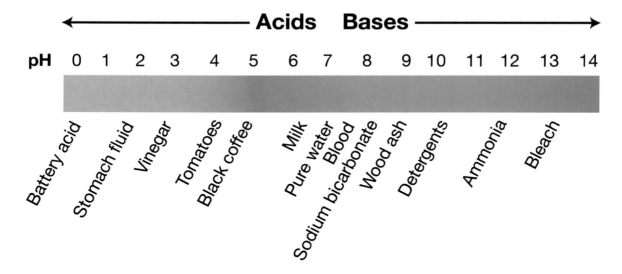

Notice that the pH for blood is near 7, close to the pH of water. Our bodies are made mostly of water, and blood carries nutrients throughout our bodies. It is important that the pH of blood be near 7 since many of our cells and tissues would be damaged if the pH were much higher or much lower than 7. However, notice that the pH for stomach fluid is even lower than the pH for vinegar. Why is stomach fluid so acidic? As it turns out, your stomach makes hydrochloric acid (HCl). This acid helps break down your food so that it can be carried to other places in your body. The inside of your stomach has a special lining that is designed to prevent the acidic stomach fluid from causing damage.

6.5 Acid-Base Indicators

Sörensen was able to create a pH scale by using a set of indicators that change color as the pH changes. Litmus paper is one type of acid-base indicator.

There are also other kinds of acid-base indicators. Some indicators change colors at very low pH, and others don't change until the pH is very high. Some indicators even change colors twice. The following chart shows a few acid-base indicators and the pH range in which they change color.

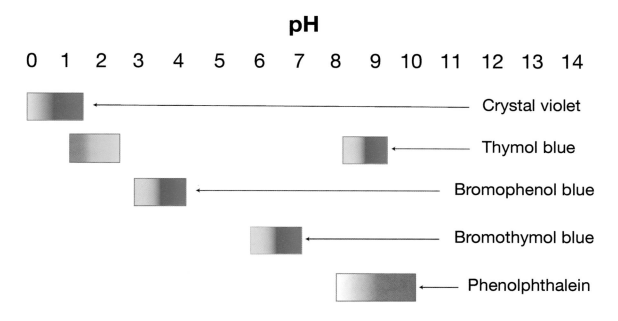

Notice that crystal violet changes colors at very low pH. Below pH 1, crystal violet is yellow, but above pH 1 it is blue. Now look at phenolphthalein. Below pH 9 phenolphthalein is not colored, but above pH 9 it turns pink. This narrow range of some pH indicators is very useful for finding the pH of a solution and also determining how concentrated the acid or base is.

In general, any molecule that changes color when the pH changes can be considered an acid-base indicator. There are many different acid-base indicators, including red cabbage juice, which is an acid-base indicator that is easy to make and fun to use.

6.6 pH Meters

Litmus paper and other indicators make it relatively easy to tell whether a solution is an acid or a base, but measuring hydrogen ion concentration and determining the strength or weakness of an acid or base can be more complicated.

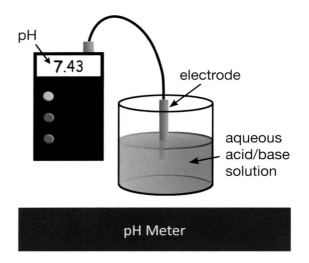

Recall from Section 6.3 that Michael Faraday discovered that acids and bases are electrolytes that conduct electricity. A pH meter works by measuring electrical resistance in a water solution (how well electricity moves through the solution). A pH meter has a probe, or electrode, that is usually made of glass and is connected to a meter that measures how much electricity an acid or base can conduct.

In 1906 two scientists, Fritz Haber (1868-1934 CE) and Zygmunt Klemensiewicz (1886-1963 CE) tried to create the first pH meter. They made glass probes and attempted to measure pH directly by inserting a probe into a solution, but the glass needed to be very thin and the probes broke easily. They were never able to get their early pH meter to work.

Chapter 6: Acids, Bases, and pH

ARNOLD BECKMAN
1900-2004 CE

In 1934 Arnold Beckman (1900-2004 CE) became the inventor of the first successful pH meter. Beckman was a chemistry professor at the California Institute of Technology (Caltech) when he was asked by the California Fruit Growers Exchange to find a way to measure the acidity of lemon juice. Members of the California Fruit Growers Exchange grew most of the citrus fruit in California at that time, and they needed a quick and easy way to see how acidic the fruit was. This information helped them decide when the fruit was ready to harvest.

After Beckman had perfected the first pH meter, he went into the business of producing and selling them. The first pH meters went on the market in 1935. Many people believed that only about 600 pH meters would be needed to supply chemistry labs around the world, but Beckman proved them wrong. Over the next couple of decades, Beckman's company grew, developing and selling other scientific instruments as well as the pH meter, and he eventually became a millionaire. Beckman was very generous with his money, contributing over 400 million dollars to science research and education during his lifetime.

Chemists now use both portable and stationary pH meters. A portable pH meter can be transported easily in a backpack or pocket and can be used outdoors to measure the pH of rivers, ponds, geothermal pools, and other water sources. Stationary pH meters, also called benchtop meters, are usually more accurate than portable pH meters and are used inside a laboratory.

6.7 Summary

- Acids are generally sour in taste, not slippery to the touch, and dissolve metals.
- Bases are generally bitter in taste, slippery to the touch, and form precipitates with metals.
- Concentration is defined as the number of units in a given volume.
- Indicators such as litmus paper can be used to determine if a solution is an acid or a base.
- The pH of a solution measures the ion concentration.
- pH can be measured by pH meters, pH paper, and acid-base indicators.

6.8 Some Things to Think About

- Why do you think it's important for chemists to know the properties of acids and bases?
- What are some other foods that you think are acidic? Why do you think so?
- Explain the difference between an Arrhenius acid and an Arrhenius base.
- Do you think that when the pH of a substance is near the extreme ends of the pH scale it is more likely to be harmful? Why or why not?
- In your chemistry lab how would it be helpful to have a variety of pH indicators?
- Why do you think the invention of the pH meter was helpful to chemists?

Chapter 7 Acid-Base Neutralization

7.1	Introduction	62
7.2	Titration	63
7.3	Plotting Data	64
7.4	Plot of an Acid-Base Titration	66
7.5	Summary	70
7.6	Some Things to Think About	71

7.1 Introduction

When an acid is added to a base, or a base is added to an acid, an acid-base reaction occurs. An acid-base reaction is a special type of exchange reaction. Recall that in an exchange reaction atoms in one molecule trade places with atoms in another molecule.

An example of an acid-base reaction is the chemical reaction between vinegar and baking soda. Vinegar (acetic acid) is an acid, and baking soda (sodium hydroxide) is a base. When vinegar and baking soda are mixed in water, one of the hydrogen atoms of the vinegar trades places with the sodium atom of the baking soda. However, there is a second step — a decomposition reaction in which the intermediate products break apart and make carbon dioxide and water. You can see that the end product of the reaction between vinegar and baking soda is a solution of water and sodium acetate (a salt), with carbon dioxide being given off as gas bubbles.

Acid-base exchange reaction

acetic acid (vinegar) + sodium bicarbonate (baking soda)

A hydrogen ion from the vinegar breaks away and replaces the sodium ion on the baking soda.

The vinegar picks up the free sodium ion and forms sodium acetate.

sodium acetate

Decomposition reaction

hydrogen ion, hydroxyl ion → water

Carbon dioxide is released as a gas and bubbles away.

The sodium acetate stays in the solution dissolved in the water.

An acid-base reaction is also called a neutralization reaction. When an equal concentration of a strong acid reacts with an equal concentration of a strong base, the resulting solution becomes neutral — neither acidic nor basic — because all the molecules of both the acid and the base have reacted with each other.

Another way of saying this is that when an acid reacts with a base, the atoms that make the acid acidic (hydrogen ions) react with the atoms that make the base basic (hydroxide ions) to form water and a salt. When this happens, the acid and base neutralize each other.

As an example of an acid being neutralized, you might think about a person who gets acid indigestion after eating too many chili cheese fries. Acid indigestion results when the stomach has produced too much acid and the excess acid causes pain. Taking an antacid can reduce the pain because an antacid is a base and when eaten will neutralize stomach acid. Antacids are not very strong (not concentrated) so they are safe to eat when the need arises.

7.2 Titration

Why do we need to take antacids that are not very concentrated? How can we be sure that the antacid we take is not too strong but just strong enough to fix our indigestion? If we have an acid or a base, how can we know whether it has a concentration that is safe for us to use?

As it turns out, chemists can determine how concentrated and how strong an acid or a base is by using a technique called a titration. The word titration comes from the old French word *titre* which means assay. An assay is a test of quality. By doing an acid-base titration, a chemist can test for the quality of an acid or base solution — how strong, weak, or concentrated it is.

The fundamental idea behind a titration is to use a solution of an acid or base of known concentration to find the concentration of an unknown acid or base solution. An acid-base indicator or a pH meter is used to observe the acid-base reaction during the titration.

To do a titration, an acid is added to a base or a base is added to an acid, and the pH of the solution is observed as it changes. If the acid or base is added a little at a time, the pH of the solution will change slowly enough to be observed by a pH meter or acid-base indicator.

For example, if the titration starts with a beaker full of vinegar and red cabbage juice acid-base indicator, the color of the liquid will be a bright pink. If one spoonful of baking soda is added, there is not enough base to completely neutralize the acid, so the solution will still be pink. However, if more spoonfuls of baking soda are added one at a time, eventually there will be enough base to neutralize the acid, and the acid-base indicator will change color.

By plotting the data on a graph, the concentration of the unknown acid or base can be determined. Before we see how this works, let's take a look at how to plot data.

An acid; no base added.

One spoonful of base added, but the solution is still acidic.

The solution starts to change color as more spoonfuls of base are added.

Eventually the solution changes color completely as the acid is neutralized.

7.3 Plotting Data

One way to examine a titration in detail is to plot or graph the data. A plot is a handy visual tool that scientists use to help them understand data. Plots can be made of almost any data. For example, you might notice that the older members in your family are usually taller than the young members, so you could say that there is a connection between age and height. A plot can be made to illustrate the relationship between age and height.

To make a plot of age vs. (versus) height, the first step is to gather the data to be plotted. For this example, the data might look something like the following.

Age	Height	
Age 1	.6 m	(2 ft)
Age 6	.9 m	(3 ft)
Age 8	1.2 m	(4 ft)
Age 11	1.5 m	(5 ft)
Age 30	1.7 m	(5.5 ft)
Age 40	1.8 m	(5.8 ft)
Age 60	1.75 m	(5.75 ft)

Once the data has been collected, the plot can be created. First, a horizontal line is drawn. This line is called the x-axis. Another line is drawn perpendicular (vertically) to the first and meets the first line at the bottom of the left-hand side. This vertical line is called the y-axis.

To plot the data in this example, the x-axis is labeled "age" and the y-axis is labeled "height." The age of the person is marked on the graph with a vertical dotted line, and the height of the person is marked with a horizontal dotted line. The point where the two lines intersect is marked with a red dot, called a point. A solid black line can then be drawn to connect all the points on the plot.

From this plot we can tell that, in general, as a person gets older, they grow taller. (The drawn black line goes up as the age goes up.) This graph also shows that a person stops growing when a certain age is reached. (The drawn black line levels off, showing that no significant growth occurs after age 20 or so.) Plotting is a tool that scientists use to organize their data in a way that makes it easier to understand.

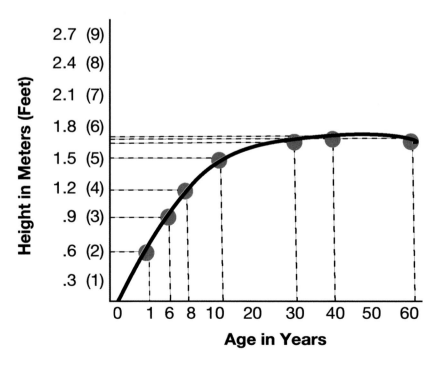

7.4 Plot of an Acid-Base Titration

Let's take a closer look at how to plot an acid-base titration and how to find out the concentration of an unknown acid or base. Imagine that we have a beaker half-full of household vinegar. We know that household vinegar is made of acetic acid molecules, but we may not know how many acetic acid molecules the beaker contains. In other words, we don't know how concentrated our household vinegar is. Imagine that we also have a box of baking soda. Using a scale we can measure the weight of the baking soda. By knowing the weight of the baking soda, we know how many molecules it has.

Wait! How can we know this? Let's use a box of ping-pong balls as an example. Imagine you have a box full of ping-pong balls. You don't know how many ping-pong balls you have, but you can find out the weight of one ping-pong ball. You can then calculate the total number of ping-pong balls. First, by dumping out all the ping-pong balls and weighing the empty box, you can get the weight of the box alone. Next, if you put all the ping-pong balls back in the box and weigh the full box, you can get the total weight. Now you can subtract the weight of the empty box from the total weight of the box with the ping-pong balls in it to find out the weight of all your ping-pong balls together.

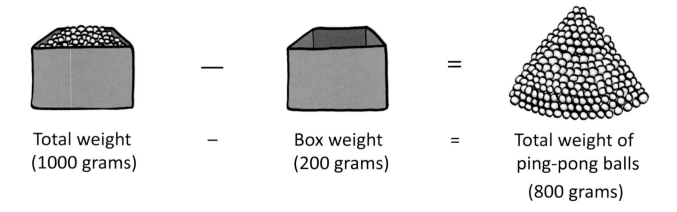

Total weight (1000 grams) − Box weight (200 grams) = Total weight of ping-pong balls (800 grams)

If you then divide the weight of all the ping-pong balls by the weight of one ping-pong ball, you can calculate how many ping-pong balls you have.

| Total weight of ping-pong balls (800 grams) | ÷ | Weight of one ping-pong ball (8 grams) | = | Number of ping-pong balls (100) |

The same is true for calculating how many molecules you have in a given amount. If you have a teaspoonful of baking soda that weighs 100 grams and you know the weight of one baking soda molecule, you can figure out the number of baking soda molecules in 100 grams.

But wait! How do we know the weight of one baking soda molecule? We can't weigh a baking soda molecule directly, but we do know that one baking soda molecule contains one carbon atom, three oxygen atoms, one sodium atom, and one hydrogen atom. Because we can use the periodic table of elements to find the atomic weight of each atom, we can add the atomic weights of all the atoms together to get the molecular weight — the weight of one molecule.

We can see that the molecular weight of sodium bicarbonate is 84 amu. But wait! What is an amu and how many grams is that? An amu is a measure of the atomic mass of an atom. Amu stands for atomic mass units. An atomic mass unit is equal to 1/12th the mass of a carbon atom.

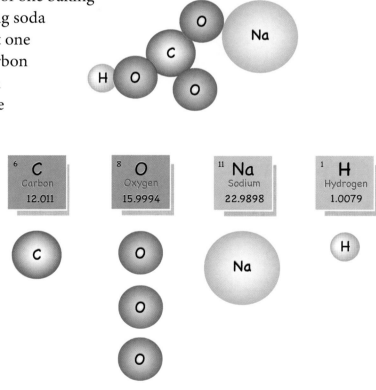

1x12 amu + (3x16 amu) + (1x23 amu) + (1x1 amu) = (84 amu)

As you can imagine, one atomic mass unit is a very small number. In fact, this number is so small that it is not useful for chemists doing chemistry calculations. Instead of using atomic mass units, chemists figured out how to measure the weight of a group of atoms. This group of atoms is called a mole. A mole is just a name that is used to represent a certain number of "things." These "things" can be atoms, molecules, ions, or even oven mitts or baseball caps. A mole is a way to count atoms and molecules just like a dozen is a way to count eggs. The only difference is that a mole is a very large number.

$$\text{One mole} = 602{,}200{,}000{,}000{,}000{,}000{,}000{,}000$$
$$(6.022 \times 10^{23})$$

A mole is 6022 followed by 20 zeros, and it's such a big number that it won't even fit on most calculators. It's so big that if you had a mole of marbles, it would be bigger than the Moon! But atoms are very tiny, so a mole of atoms is a nice, manageable size. A mole of most atoms will fit in the palm of your hand.

AMEDEO AVOGADRO
1776-1856 CE

This big number for one mole is called Avogadro's constant and is named after Italian scientist Amedeo Avogadro who had the idea that the number of gas molecules in a given volume is the same no matter what kind of gas it is. Based on his idea, Avogadro was able to calculate the number of molecules in the given volume. Today, Avogadro's number is used to relate the number of atoms and molecules to their atomic and molecular weights.

By definition, one mole of carbon atoms equals 12 grams, one mole of hydrogen atoms equals 1 gram, and 1 mole of sodium bicarbonate molecules equals 84 grams. This comes in very handy for acid-base reactions. Instead of worrying about the number of molecules needed to neutralize a reaction, we can use the number of moles. One mole of sodium bicarbonate will neutralize one mole of vinegar. We know that one mole of baking soda weighs 84 grams and we can measure this on a scale!

Now that we know that 84 grams of baking soda equals one mole, we can use a titration to find out how many moles of acid are in a solution. Let's go back to the titration of vinegar and baking soda. We have a beaker half-full of vinegar and we have a box of baking soda. Using a scale we can measure how much a teaspoonful of baking soda weighs in grams. We also know that one mole of baking soda will neutralize one mole of vinegar.

By using a pH meter or an acid-base indicator, we can see the pH of the solution change as we slowly add baking soda to the vinegar. We can record the pH change for each teaspoon of baking soda added and then plot a graph from this data. The graph may look something like the one following.

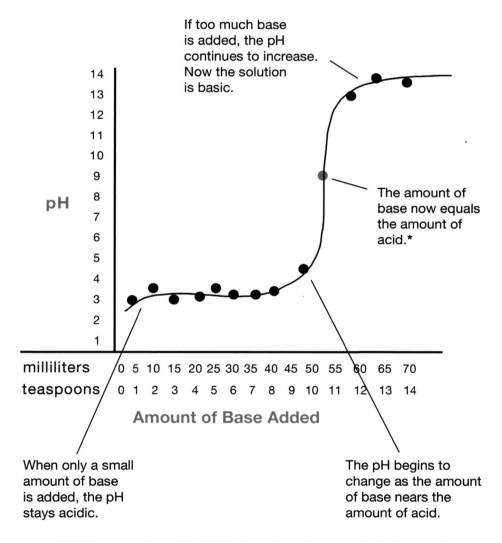

* In this example the pH is slightly higher than 7 at the neutralization point because the product of the reaction, sodium acetate, is slightly basic.

The x-axis (horizontal) is labeled *Amount of Base Added*, and the y-axis (vertical) is labeled *pH* but could also be labeled *Color*. Notice that the graph looks like a snake with two curved sections. The midpoint between the two curved sections is the point where the amount of base equals the amount of acid. If you know how many moles of base are in a teaspoonful and you know how many teaspoonfuls were added, you know how many moles of base it takes to neutralize *all* the acid. Therefore, you know how many moles of acid were in the beaker you started with!

You Do it!

1. If it takes 84 grams of baking soda to neutralize a beaker of acetic acid, how many moles of acetic acid do you have?

2. If if takes 42 grams of baking soda to neutralize a beaker of acetic acid, how many moles of acetic acid do you have?

3. If if takes 168 grams of baking soda to neutralize a beaker of acetic acid, how many moles of acetic acid do you have?

(See end of chapter for solutions.)

7.5 Summary

- An acid-base reaction is a special type of exchange reaction.

- An acid and a base neutralize each other in an acid-base reaction.

- Equal amounts of acid and base completely neutralize each other.

- Plotting data can make it easier to understand.

- A mole is the name for a certain number of atoms, molecules, or ions.

- The concentration of an unknown acid or base can be found using a titration.

7.6 Some Things to Think About

○ Explain how an acid-base reaction occurs.

○ Why do you think a chemist would use an acid-base indicator or a pH meter during a titration?

○ What do you think the results of a titration would tell a chemist?

○ Review the chart of data and the plot in Section 7.3 and compare how easy or difficult it is to interpret and read data on each. Do you think there are times when you would want to use a chart and times when using a plot would be better? Why?

When do you think you might want to use both? Why?

(You Do It! solutions on next page.)

You Do It! – Solutions

Acid-Base Neutralization

We know that 84 grams of baking soda equals one mole (pages 67-68).

We also know that one mole of baking soda neutralizes 1 mole of acetic acid (page 68). Looking at the illustration on page 62, we can see that there are the same kind and number of atoms in the molecules before the reaction and in the new molecules that result from the reaction. We can write the reaction as:

$$CH_3COOH + NaCO_3H \longrightarrow NaCH_3COO + CO_2 + H_2O$$

acetic acid + baking soda   sodium acetate + carbon dioxide + water

Because we know how the atoms and molecules behave during the reaction, we can see that 2 moles of baking soda will neutralize 2 moles of acetic acid, 3 moles of baking soda will neutralize 3 moles of acetic acid, 0.5 moles of baking soda will neutralize 0.5 moles of acetic acid, and so on.

The questions in this *You Do It* section ask how many moles of acid are neutralized by 84 grams, 42 grams, and 168 grams of baking soda. In each case all we need to do to solve the problem is to convert grams of baking soda to moles of baking soda. Since an equal concentration of baking soda will neutralize an equal concentration of acetic acid, we know that the number of moles of baking soda and the number of moles of acetic acid will be the same.

To convert grams to moles we use a "conversion factor." A conversion factor states mathematically the relationship between two quantities. For baking soda we can write our conversion factor as:

$$\frac{1 \text{ mole}}{84 \text{ grams}} \quad \text{or} \quad \frac{84 \text{ grams}}{1 \text{ mole}}$$

(Continued on next page.)

You Do It! – Solutions (cont.)

Acid-Base Neutralization (cont.)

To solve these problems, multiply the number of grams by the conversion factor.

1. If it takes 84 grams of baking soda to neutralize a beaker of acetic acid, how many moles of acetic acid do you have?

$$\cancel{84 \text{ grams}} \times \frac{1 \text{ mole}}{\cancel{84 \text{ grams}}} = 1 \text{ mole}$$

There is 1 mole of acetic acid that is neutralized by 84 grams of baking soda.

2. If it takes 42 grams of baking soda to neutralize a beaker of acetic acid, how many moles of acetic acid do you have?

$$42 \text{ grams} \times \frac{1 \text{ mole}}{84 \text{ grams}} = 0.5 \text{ mole}$$

0.5 mole of acetic acid is neutralized by 42 grams of baking soda.

3. If it takes 168 grams of baking soda to neutralize a beaker of acetic acid, how many moles of acetic acid do you have?

$$168 \text{ grams} \times \frac{1 \text{ mole}}{84 \text{ grams}} = 2 \text{ moles}$$

2 moles of acetic acid are neutralized by 168 grams of baking soda.

Chapter 8 Nutritional Chemistry

8.1 Introduction — 75

8.2 Minerals — 76

8.3 Vitamins — 78

8.4 Carbohydrates — 79

8.5 Starches — 80

8.6 Cellulose — 82

8.7 Summary — 83

8.8 Some Things to Think About — 83

8.1 Introduction

We saw in the last chapter how acid-base reactions can help neutralize stomach acid, but what other reactions occur in the body? Where do we get the molecules we need to live? What happens to the food we eat? We may know from experience that when we skip a meal or are unable to eat because of illness, we become weak and lack energy. Our bodies require food to help us grow and keep our "engines running." Without food we would not survive.

Unlike plants, we cannot stick our feet in the soil, lift our hands to the Sun, and make our own food. In fact, we rely on plants and animals to provide the food our bodies need to keep us going. But what is in food and what does food do for us?

There are many different kinds of food the body requires to stay healthy. These foods provide important nutrients including vitamins, proteins, fats, and carbohydrates. We get these important nutrients from eating a variety of foods. The nutrients we get from eating foods provide the necessary molecules our bodies need to grow and function properly. Vitamins, like those found in carrots, help our eyes work. Fats found in vegetable oil and butter help our brain and other tissues function. Proteins from fish and meats help our bones heal and our muscles grow. Carbohydrates, like those found in bread, potatoes, and sweets, provide us with energy.

8.2 Minerals

The smallest nutrients found in many of the foods we eat are minerals. Minerals are salts of various elements. Minerals are not manufactured inside living things but are found in the Earth's soil. Plants get their minerals directly from the soil, and animals get most of their minerals from plants or other animals. Humans require a modest amount of seven different elements found in minerals (calcium, phosphorus, potassium, chlorine, sodium, magnesium, and sulfur) and extremely small, or trace, amounts of several other elements, such as fluorine, cobalt, copper, and iron.

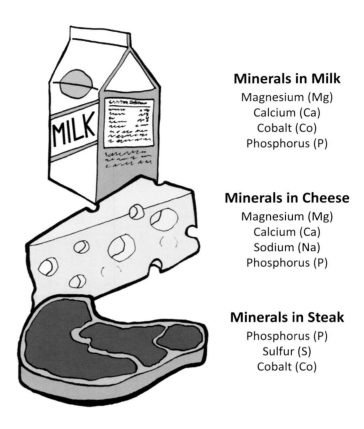

Minerals in Milk
Magnesium (Mg)
Calcium (Ca)
Cobalt (Co)
Phosphorus (P)

Minerals in Cheese
Magnesium (Mg)
Calcium (Ca)
Sodium (Na)
Phosphorus (P)

Minerals in Steak
Phosphorus (P)
Sulfur (S)
Cobalt (Co)

Elements such as calcium (Ca), magnesium (Mg), phosphorus (P), sulfur (S), and cobalt (Co) are found in foods such as cheese, milk, and meat.

Minerals are not used for fuel but are used to help maintain many of the systems inside our cells. For example, calcium, magnesium, and phosphorus help strengthen bones and harden teeth; iron is essential for oxygen binding in blood cells; and copper is important for the correct functioning of nerve tissues.

We get the different minerals we need from a variety of different foods. Calcium and phosphorus are found in milk products, leafy vegetables, and egg yolks. Sodium, potassium, and chlorine are found in table salts, cheese, and dried apricots. Magnesium is found in whole-grain cereals and a variety of leafy green vegetables. Iron is found in meat, dried nuts, and molasses; and zinc is found in seafood, nuts, and yeast. Without minerals our bodies would not function properly. The lack of necessary minerals can result in a variety of diseases or even death.

Below is a table showing some of the minerals our bodies need, what foods they are found in, and how the body uses them.

Mineral	Sources	Uses in Body
Potassium (K)	avocados, apricots, meats, leafy greens	muscle contraction, making proteins
Sodium (Na)	table salt, cheese, cured meats, celery	maintains water balance and cellular pumps
Calcium (Ca)	milk, milk products, egg yolk, leafy greens	needed for bones and teeth, blood clotting, nerve impulses
Chlorine (Cl)	table salt, seaweed, tomatoes, lettuce	activates salivary amylase, helps transport CO_2
Magnesium (Mg)	milk, dairy products, whole grain cereals, nuts, spinach	required for normal muscle and nerve function, bones
Sulfur (S)	meat, milk, eggs, legumes, garlic, onions	found in some proteins, needed for cartilage, bones, and tendons
Phosphorus (P)	milk, eggs, meat, fish, nuts, whole grains	needed for bones and teeth, nerve activity, and energy storage
Cobalt (Co)	liver, lean meat, fish, milk, broccoli	needed for vitamin B_{12}
Fluorine (Fl)	fluoridated water	tooth structure, prevents dental cavities, and may prevent osteoporosis
Copper (Cu)	liver, shellfish, whole grains, meat, beans, nuts, potatoes	needed for the manufacture of nerve tissues and blood molecules
Iodine (I)	iodized salt, shellfish, cod liver oil, milk, eggs, seaweed, potatoes	needed for making thyroid molecules
Manganese (Mn)	nuts, whole grains, fruit, leafy vegetables	needed for making fats, blood molecules, and carbohydrates
Iron (Fe)	meat, liver, nuts, egg yolk, molasses, beans, spinach	essential part of blood molecules that bind oxygen
Selenium (Se)	seafood, meat, cereal, milk	component of certain proteins
Zinc (Zn)	seafood, nuts, yeast, cereal, meat, beans	component of some proteins, required for wound healing, taste, and smell
Chromium (Cr)	liver, meat, broccoli, yeast, fruit, whole grains	component of some proteins and needed for glucose use

8.3 Vitamins

Vitamins are another set of molecules needed for a healthy body. Humans require a variety of vitamins for growth and good health. Like minerals, vitamins are not used as sources of energy but instead function as "helper" molecules for several different chemical reactions that occur in the body. Like minerals, most vitamins are found in the food we eat. However, human bodies do make two vitamins — vitamin D which is made in our skin and vitamin K which is made by beneficial *E. coli* bacteria that live in our intestines. Because there is no one food that contains all of the vitamins necessary for the healthy maintenance of the body, it is important to eat a variety of foods.

Below is a table showing some of the vitamins we need, what foods they are in, and how the body uses them.

Vitamin	Sources	Uses in Body
Fat-soluble		
Vitamin A	yellow and green vegetables, fish liver oil	needed for normal tooth and bone development, eyes, skin
Vitamin E	wheat germ, vegetable oil, dark green vegetables, nuts	protects cell membranes and prevents hardening of arteries
Vitamin D	produced in the skin by ultraviolet light, also in cod liver oil, egg yolk, milk	needed for normal tooth and bone development, blood clotting
Vitamin K	made by bacteria inside the body, also in cabbage, liver, green leafy vegetables	needed to form some proteins and for cell function
Water-soluble		
Vitamin C	fruits, vegetables, tomatoes, potatoes	used in the formation of all connective tissue, helps iron absorption
Vitamin B_1	lentils, green peas, pork, whole grains	needed for making certain sugars
Vitamin B_2	meats, milk, egg white, green leafy vegetables, legumes	needed for some protein function, making red blood cells
Vitamin B_{12}	liver, meat, fish, dairy foods, eggs	needed for the nervous system, bone marrow, and making DNA
Vitamin B_6	liver, fish, bananas, sweet potatoes	needed for making DNA and certain proteins
Vitamin B_5	liver, eggs, meat, legumes, potatoes, milk	needed to make fats, steroids, and blood molecules

Some of the vitamins we need are soluble (will dissolve) in water, and the body absorbs these vitamins directly when the digestive tract absorbs water. However, some vitamins are soluble only in fat, and the body absorbs these vitamins when we eat fats. For example, vitamin A is an important vitamin that is only soluble in fats, so it is important to eat enough fats for the body to get the vitamin A it needs. Fat-soluble vitamins are stored by the body but water-soluble vitamins are not. Our body eliminates excess water-soluble vitamins, but because our bodies store fat-soluble vitamins, it is possible to get too much of these. For example, too much vitamin A is toxic and can cause nausea, vomiting, or bone and joint pain.

8.4 Carbohydrates

Carbohydrates are another set of molecules essential for living things. Carbohydrates are the most abundant class of biological molecules and are found in every living thing. The word carbohydrate comes from the name of the element carbon and the Greek word *hydor*, meaning "water," so a carbohydrate is a molecule made of both carbon and water.

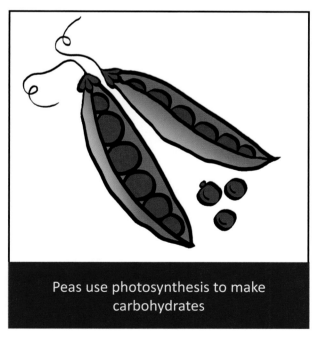

Peas use photosynthesis to make carbohydrates

Carbohydrates are made inside living things through two main biochemical processes: gluconeogenesis and photosynthesis. Gluconeogenesis comes from the Greek words *glykys* which means "sweet," *neo* which means "new," and *gen* which means "birth" or "produce." Gluconeogenesis literally means the "new production of sweet molecules." Gluconeogenesis occurs in the liver and kidneys of humans, and it is the biochemical pathway that makes carbohydrates when no food is consumed (for example, during a fast). Photosynthesis, on the other hand, is the biochemical process that plants use to convert light energy into food energy, or sugars. The bulk of carbohydrate molecules come from photosynthesis.

The simplest carbohydrates are the sugars. Sugars are relatively small molecules. They taste sweet and can be easily broken down by the body to provide quick energy. The smallest carbohydrates are called monosaccharides. *Mono-* is a Greek prefix meaning "one," and saccharide comes from the Greek word *sakcharon*, meaning sugar. A monosaccharide is "one sugar." The single sugars glucose and fructose are monosaccharides.

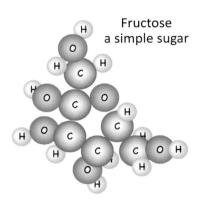
Fructose
a simple sugar

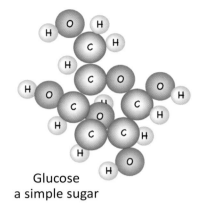
Glucose
a simple sugar

Sucrose is an example of a disaccharide, which is a molecule made of two single sugars. Sucrose contains a molecule of glucose connected by a chemical bond to a molecule of fructose.

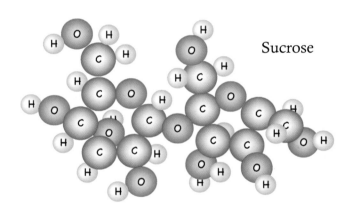
Sucrose

Sucrose is common table sugar and is the same sugar we buy in the store and put on strawberries.

8.5 Starches

When more than a few saccharides, or sugars, are hooked together, the molecule is called a polysaccharide. *Poly* means "many," and a polysaccharide is made of "many sugars." Polysaccharide molecules usually contain ten or more monosaccharides.

There are two general types of polysaccharides — starch and cellulose. Starches are the molecules that provide our bodies with most of the energy we need in order to live and work. Potatoes, pasta, and bread are excellent sources of starches.

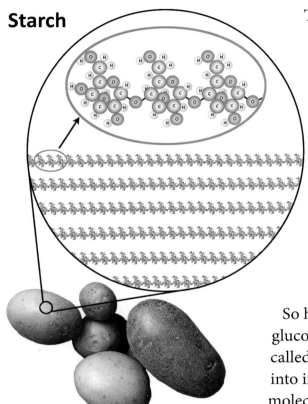

Starch

Potatoes
Courtesy of USDA/ARS/
by Scott Bauer

There are three main kinds of starches. Glycogen is a starch that animals produce in their livers and store in their muscles. Amylose and amylopectin are two starches that are made by plants and are the main energy storage molecules found in rice and potatoes.

All of these polysaccharides are composed entirely of glucose molecules linked together to make long chains. They can have as many as 3,000 glucose units hooked together in a row.

So how do our bodies use these long chains of glucose for energy? Our bodies use special proteins called enzymes to break the long chains of glucose into individual glucose molecules. The single glucose molecules are used directly by the body for energy. But if it is only the glucose our bodies need, why not eat only the simple sugars and have a diet rich in candy-coated sugar bomb cereals?

If we ate only simple sugars, our bodies would use up all of the energy in these molecules too quickly, leaving us feeling tired. The long chains in polysaccharides provide "storage" for the energy molecules so the body can use these over a longer period of time, giving us enough energy to ride bikes, swim, or run.

8.6 Cellulose

Cellulose differs from the starches only in how the glucose molecules are hooked together. The links between the glucose molecules in cellulose are different from the links between those of starches. For the starches, the oxygen atom that connects the two glucose molecules is pointing down. However, the oxygen between the two glucose molecules in cellulose is pointing up. The direction of this bond is the only difference between these two molecules, but it makes a huge difference to us.

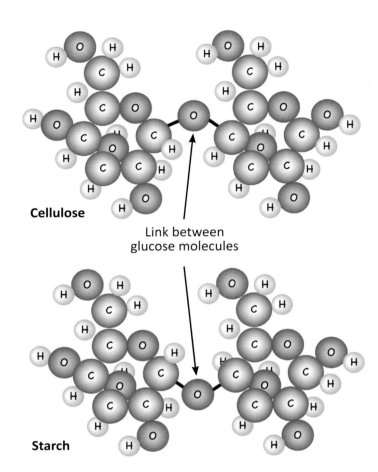

Cellulose is the main ingredient of wood, cotton, flax, wood pulp, and other plant fibers. It is even in grass. However, wood and grass are not main staples of our diet and are not served for Sunday brunch. In fact, although cellulose has the same glucose molecules that starches have, humans cannot use cellulose for food energy at all.

Many animals, including humans, do not have the enzyme required to break the bonds between the glucose molecules in cellulose. Our enzymes only recognize the bonds in the starches, so we cannot graze the lawn for breakfast. Some animals, like cattle and horses, have bacteria in their digestive system that provide the necessary enzymes to break cellulose links, so these animals can eat grass for food energy, but we cannot.

8.7 Summary

- Some of the nutrients our bodies require to stay healthy are vitamins, proteins, fats, and carbohydrates.

- Carbohydrates are molecules that give our bodies energy.

- Simple carbohydrates are sugars, and larger carbohydrates are the starches and cellulose.

- Long chains of polysaccharides store energy for the body. This energy is released when the body breaks down the chains into individual glucose molecules.

- Our bodies cannot use cellulose for energy because they cannot break the bonds in a cellulose molecule. Animals such as horses and cows can eat grass for food because they have bacteria that provide the enzymes needed to break the bonds in cellulose molecules.

8.8 Some Things to Think About

- Why do you think learning about nutrients can help us stay healthy?

- For one day, record all the foods you eat. Look up the foods on the charts in this chapter and record the nutrients you have received. You can also do research in the library or on the internet to find out more.

- Why is it important to eat a variety of foods? Which foods would you include in your ideal diet?

- Do you think your body is getting sugars when you eat plants ? Why or why not?

- How do you think table sugar is made?

- Why do you think your body needs starches?

- Where do starches come from?

- What do you think would happen if we did not have enzymes to break down starches?

- Do you think a scientist could make an enzyme we could take so we could eat grass?

Chapter 9 Pure Substances and Mixtures

9.1 Introduction — 85

9.2 Pure Substances: Elements and Compounds — 85

9.3 What Is a Mixture? — 87

9.4 Types of Mixtures — 88

9.5 Solubility of Solutions — 91

9.6 Surfactants — 94

9.7 Principles of Separation — 96

9.8 Techniques of Separation — 100

9.9 Summary — 105

9.10 Some Things to Think About — 105

Chapter 9: Pure Substances and Mixtures

9.1 Introduction

Everything you can see, taste, and touch is made of different kinds of atoms that collectively are called matter. Matter can be classified into two categories: pure substances and mixtures. In the following chart you can see that pure substances can be further divided into compounds and elements (atoms). Compounds are made of atoms, and atoms can be divided into smaller particles called protons, neutrons, and electrons. Protons and neutrons are grouped in the nucleus of an atom.

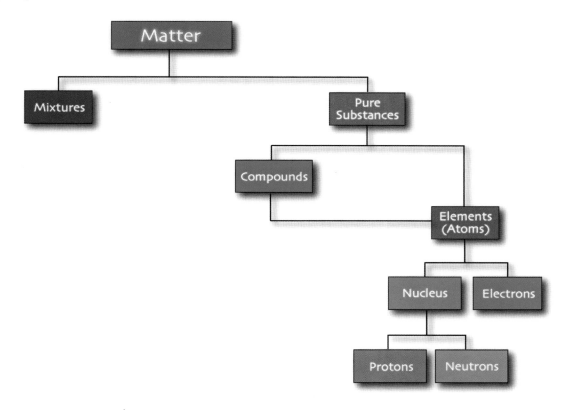

9.2 Pure Substances—Elements and Compounds

A pure substance can be either an element (atom) or a compound. An element is distinguished by its atomic number and cannot be broken down into simpler substances. A compound is made of two or more elements. Some pure substances are made of only one kind of element. For example, pure gold contains *only* gold atoms and nothing else. Likewise, pure graphite, such as the graphite in your pencil, contains *only* carbon atoms and nothing else. All pure metals such as aluminum, iron, and copper contain *only* one kind of atom. Pure oxygen gas is made of *only* oxygen atoms, and pure nitrogen gas is made of *only* nitrogen atoms.

We call such substances elemental, as in elemental gold or elemental carbon, to indicate that they are composed of only one kind of atom.

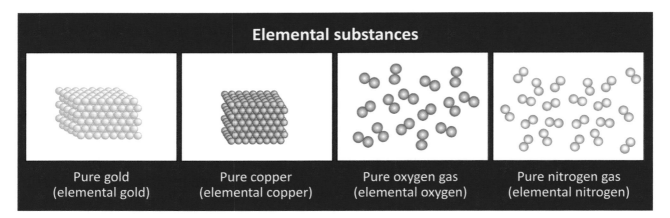

Other pure substances contain more than one element but are composed of only one type of molecule. We call these pure substances compounds. Pure water, for example, is made of two different elements: hydrogen and oxygen. Pure ammonia contains three hydrogen atoms and one nitrogen atom, and pure carbon dioxide contains two oxygen atoms and one carbon atom. Even though compounds contain more than one kind of atom, they are considered to be pure substances because they are composed of only one kind of molecule.

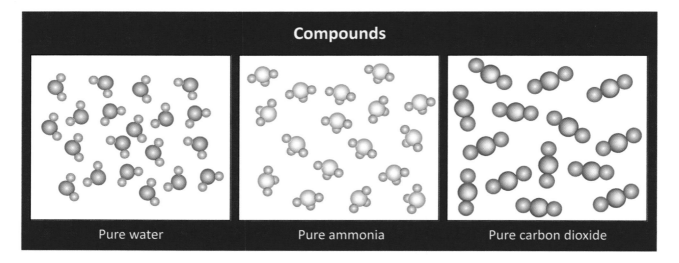

In general, a compound is *two or more atoms bonded together in a fixed ratio*. (For example, there is always one oxygen atom to two hydrogen atoms in a water molecule for a ratio of one to two.) Because the atoms in compounds are all bonded together in the same way, they are also considered pure substances.

Chapter 9: Pure Substances and Mixtures 87

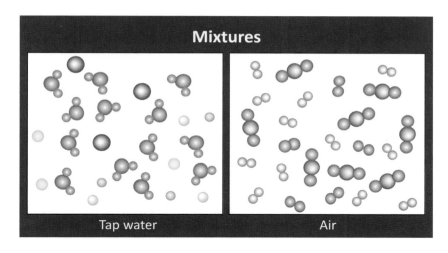

Matter can also exist in mixtures. A mixture is defined as two or more molecules physically combined but not chemically bonded. The air we breathe is a mixture of nitrogen molecules (N_2), oxygen molecules (O_2), and trace amounts of other molecules. Tap water is a mixture of water molecules and small amounts of metals and often chlorine. Unlike a compound, a mixture contains more than one type of molecule. Therefore, mixtures are *not* pure substances.

9.3 What Is a Mixture?

When we look closely at the world around us, we find that almost everything is made of a mixture. For example, as mentioned above, the air we breathe is really a mixture of several different gases, including oxygen (O_2), nitrogen (N_2), and carbon dioxide (CO_2). Notice that carbon dioxide *is not* a mixture because the different atoms (carbon and oxygen) are

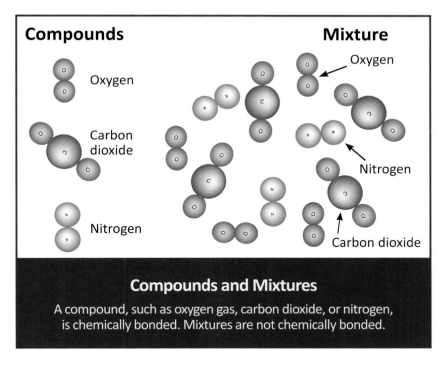

bonded chemically. Therefore, it is a compound. But air *is* a mixture because the oxygen, nitrogen, and carbon dioxide are not chemically combined, only physically mixed together. The water we drink is usually a mixture of water molecules and dissolved ions such as Na^+, Mg^{2+}, or Ca^{2+}.

Mixture of Mixtures

The foods we eat are mixtures. Bread, cheese, lasagna, and even chocolate bars are all mixtures. Looking closely, we find that mixtures are often mixtures of mixtures.

For example, bread contains yeast, oil, sugar, salt, wheat, and eggs. But each of these items is itself a mixture of molecules. Yeast is made of carbohydrates and proteins; vegetable oil has oleic acid and perhaps linoleic acid; eggs have fats, water, and protein; and wheat has many compounds, including starches, amino acids, and proteins. Most of the things around us are complex mixtures of one kind or another.

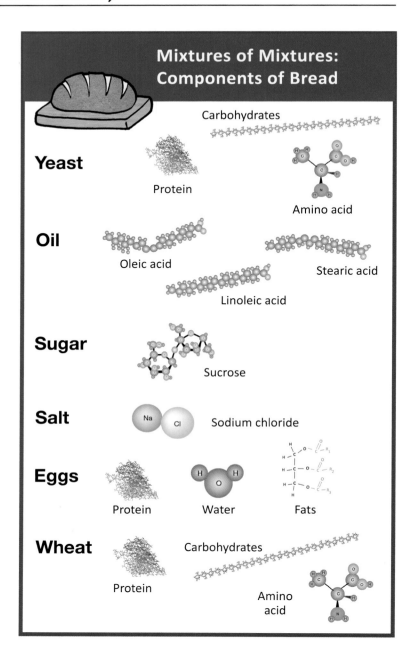

9.4 Types of Mixtures

Homogeneous and Heterogeneous

There are two main types of mixtures: homogeneous mixtures and heterogeneous mixtures. The word homogeneous means "same kind" and the word heterogeneous means "other kind." Therefore, a homogeneous mixture is a mixture that is the same throughout, and a

Chapter 9: Pure Substances and Mixtures   89

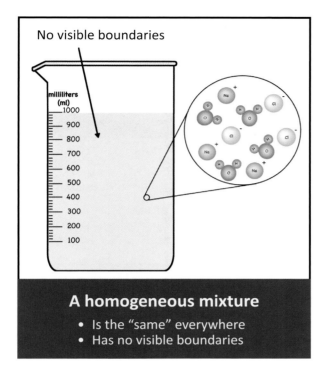

A homogeneous mixture
- Is the "same" everywhere
- Has no visible boundaries

heterogeneous mixture is a mixture that is not the same throughout. To put it more simply, a homogeneous mixture appears uniform, whereas a heterogeneous mixture is milky or even lumpy.

The main difference between a homogeneous mixture and a heterogeneous mixture is the sizes of the things that are mixed. In a homogeneous mixture, the molecules are mixed on a *molecular level* so they are essentially *invisible*. For example, the individual chlorine ions, sodium ions, and water molecules in a homogeneous mixture of salt water cannot be seen with our eyes. The molecules are too small. We do not observe a visible boundary between the sodium ions, chloride ions, and water molecules because ions and molecules are mixed.

On the other hand, a heterogeneous mixture typically has particles that are small, but much larger than individual molecules. They are on a macromolecular scale and are often visible. For example, salad dressing made of oil and vinegar (and hopefully a touch of garlic) has a visible boundary. When the dressing is shaken vigorously, the droplets become very small and the mixture turns milky, but it never becomes clear. Oil and water never mix at the molecular level. Because the atoms and molecules form macromolecular structures (oil droplets), the particles are usually large enough to see with our eyes.

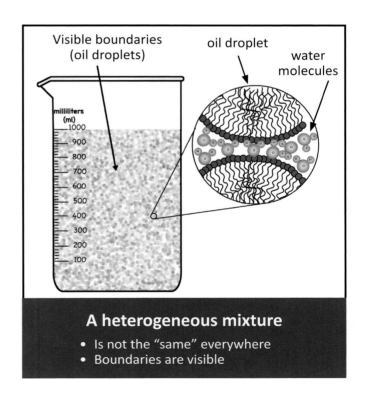

A heterogeneous mixture
- Is not the "same" everywhere
- Boundaries are visible

Solutions and Colloids

Solutions are a type of homogeneous mixture. We normally think of solutions as liquids, but in fact, the term solution can also be applied to both solids and gases. For example, 12 karat gold is a solid solution of gold, silver, zinc, and copper atoms mixed together in a solid. Air is a gaseous solution of many gases, including oxygen, nitrogen, and carbon dioxide. Soda pop is a liquid solution that contains carbon dioxide, water, and phosphoric acid. So a solution can be in any form: solid, liquid, or gas. We will learn more about solutions in the next section.

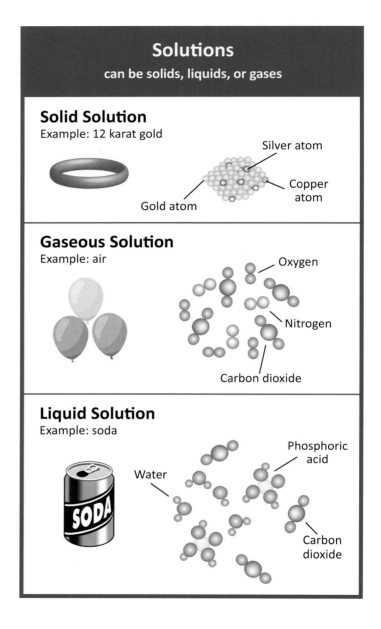

Some mixtures might seem like homogeneous solutions, but are, in fact, heterogeneous colloids. Recall that a heterogeneous mixture is characterized by the fact that it often has visible particles. A colloid, however, has particles that are quite difficult to see individually, but are still much larger than individual molecules. For example, milk is a colloid. Milk appears to the unaided eye to be a homogeneous mixture since the white color of milk is uniform throughout. However, milk is actually made up of water, proteins, and fats. The proteins and fats are gathered into oil-like particles that do not mix homogeneously with the water. So, even though the particles in milk can't be seen, milk is a heterogeneous mixture.

Chapter 9: Pure Substances and Mixtures   91

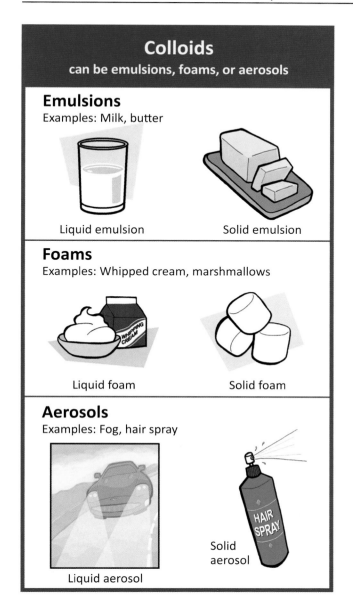

There are different types of colloids. The differences between colloids depend on whether or not the components that are mixed are solids, liquids, or gases. Milk—a type of colloid—is called an emulsion. An emulsion is a liquid (fats) mixed into a liquid (water). Butter is also a colloid and is called a solid emulsion. A solid emulsion is a liquid (proteins) mixed into a solid (fats at room temperature). Whipped cream is called a foam. A foam is a gas (air) mixed into a liquid (water and proteins). A marshmallow is a solid foam. A solid foam is a gas (air) mixed into a solid (cooked sugar at room temperature). Hair spray, fog, and smoke are aerosols. Aerosols are either liquids (water—fog) or solids (particles—smoke) mixed into a gas (air).

9.5 Solubility of Solutions

What happens when salt or sugar dissolves in water? Why do they form a homogeneous mixture? Why doesn't water dissolve in oil, or oil in water? Why do oil and water form a heterogeneous mixture? If oil does not dissolve in water, what will oil dissolve in?

These questions address the physical property called solubility. When a molecule or compound dissolves in something, we say it is soluble. That is, we will get a homogeneous mixture of atoms, molecules, or ions dispersed in each other with no clumping, droplets, or large particles. Solubility is a physical property and not a chemical property since no chemical reaction takes place. Soluble compounds form homogeneous mixtures, but insoluble compounds form heterogeneous mixtures. This is because the molecules stay in clumps or droplets and do not disperse. But what makes a compound soluble or insoluble?

Solubility and Polarity

A solution is usually made up of a small amount of one substance dissolved into a large amount of another substance. The substance that dissolves is called the solute, and the substance it dissolves into is called the solvent. The solvent is the most abundant substance. For example, a small amount of salt dissolves in a larger amount of water, so the salt is the solute and the water is the solvent. The solubility of a solute is the maximum amount of solute (in grams or moles) that dissolves in a given volume of solvent (in liters or milliliters) at a given temperature. For example, the solubility of NaCl in water is 39.12 g/100 ml at 100° C. This means that, at most, 39.12 grams of salt will dissolve in 100 ml of water at 100° C.

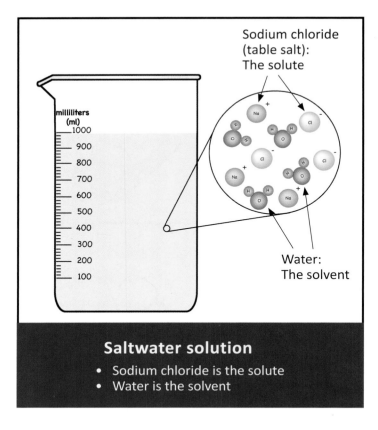

Saltwater solution
- Sodium chloride is the solute
- Water is the solvent

Perhaps the most important characteristic that determines whether a solute will dissolve in a given solvent is called polarity. A molecule that has poles with opposite electric charges is said to have polarity, or to be polar.

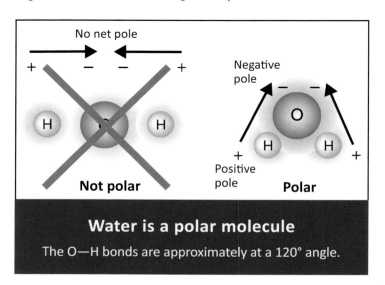

Water is a polar molecule
The O—H bonds are approximately at a 120° angle.

Polarity and Shape

For a molecule to be polar, the shape of the molecule matters. For example, water is a bent molecule with the hydrogens sticking out away from the oxygen at an angle of about 120°. If water were linear—with the hydrogens sticking out on both ends—water would not be polar. This is because the little poles (or dipoles) on each bond cancel each other out in a linear molecule, resulting in no overall charge on the molecule. However, because water is not a linear molecule, but bent, the pulling of the electrons toward the oxygen atom causes the water molecule to have a net negative charge around the oxygen and a net positive charge around the hydrogens. Water is an example of a polar molecule—it has a positively charged pole and a negatively charged pole.

Like Dissolves Like

The rule for solubility is:

Like dissolves like.

This means that polar and ionic compounds tend to dissolve in polar solvents, and nonpolar (or weakly polar) molecules tend to dissolve in nonpolar (or weakly polar) solvents. Water is a very polar solvent, so only polar (or ionic) molecules dissolve in water.

This is why sodium chloride (table salt) dissolves in water. Sodium chloride forms an ionic bond, and so it readily dissolves in a polar solvent like water.

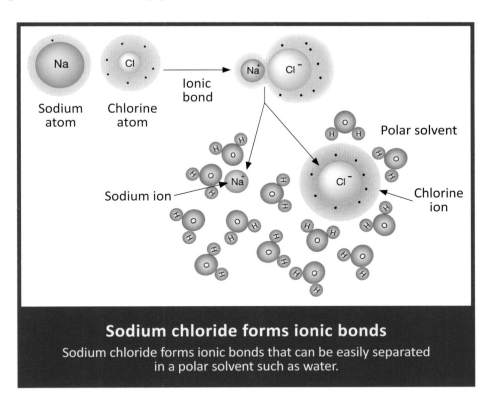

Sodium chloride forms ionic bonds
Sodium chloride forms ionic bonds that can be easily separated in a polar solvent such as water.

On the other hand, octane, a very nonpolar molecule in gasoline, will not dissolve ionic or polar molecules, but will dissolve nonpolar molecules. Some cleaners use nonpolar solvents to dissolve nonpolar substances such as glue, gum, or grease.

9.6 Surfactants

Polar substances are often called hydrophilic substances. Hydro comes from the Greek word *hydro* which means "water" and philic comes from the Greek word *philein* which means "loving." So hydrophilic literally means "water loving." Hydrophilic molecules are molecules that "love," and hence dissolve in, water.

Salts, acids and bases, alcohols, and sugars are all hydrophilic molecules. They are hydrophilic because they are either ionic, polar, or have one or more polar groups attached to them.

Nonpolar substances are called hydrophobic. Phobic comes from the Greek word *phobos* which means "to fear." So hydrophobic literally means "to fear water." Hydrophobic molecules do not like water, and so they do not dissolve in water.

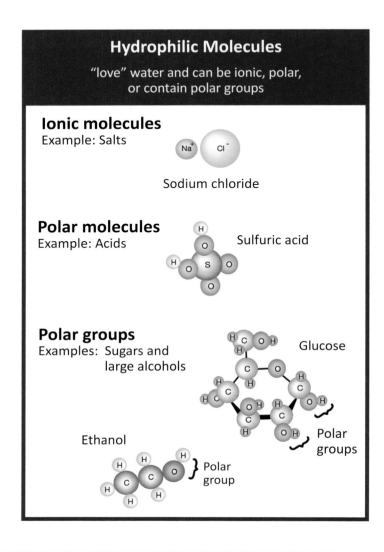

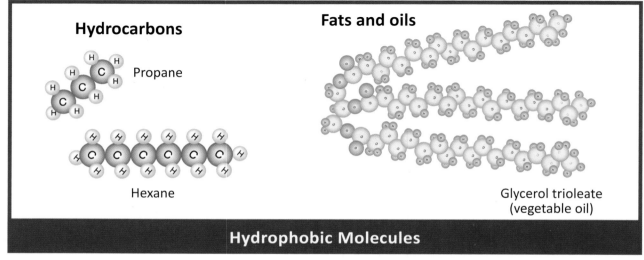

Chapter 9: Pure Substances and Mixtures

Oils, fats, and hydrocarbons such as gasoline are hydrophobic molecules because they are neither polar molecules nor do they have polar groups attached to them.

Soaps

What about molecules that are *both* hydrophilic *and* hydrophobic? Soaps are molecules that have both a hydrophobic group (tail) and a hydrophilic group (head). Soaps are part of a broader category of molecules called surfactants. Both soaps and detergents are surfactants. Surfactants are used to clean grease, oil, and other hydrophobic molecules from clothing, hands, or other items.

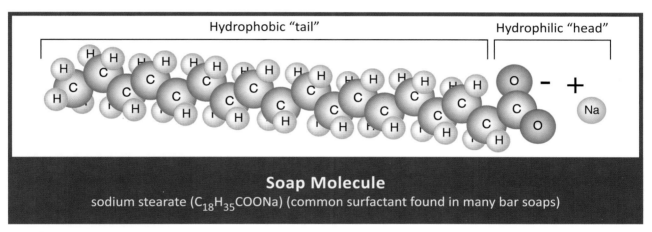

Soap Molecule
sodium stearate ($C_{18}H_{35}COONa$) (common surfactant found in many bar soaps)

Surfactants can make even nonpolar, hydrophobic molecules "dissolve" in water. That is why dish soap can clean greasy dishes and hands. Surfactants work by making an emulsion with hydrophobic molecules in the form of a micelle. For example, when a surfactant meets both water and oil, it forms a ball with the hydrophobic molecules (oil) surrounded by the surfactant. The surfactant

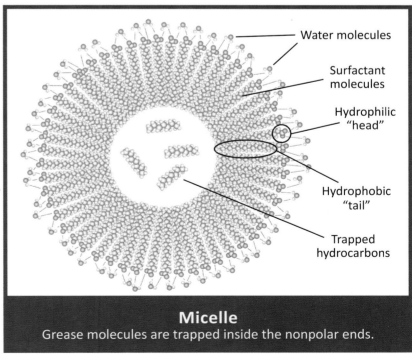

Micelle
Grease molecules are trapped inside the nonpolar ends.

molecules have their greasy tails pointed inward dissolved in the oil, and their polar heads pointed outward dissolved in the water.

Because surfactants have both a hydrophobic tail and a hydrophilic head, they are able to trap hydrophobic molecules in micelles and bring them into an emulsion. Soap solutions are colloids and are milky in appearance.

9.7 Principles of Separation

Once two items have been mixed, how can they be separated? We know how to separate pebbles and stones, but is it possible to separate smaller items like atoms and molecules? What chemical or physical techniques could help or hinder separating small items, like atoms or molecules?

The techniques used to separate mixtures depend on the kinds of items that have been mixed. For example, we can hand sort mixtures of large components, such as pebbles and stones. Hand sorting is a technique used to separate large items, and it is something everyone has done at one time or another (e.g., cleaning your room). But what about a mixture of sand and salt, or salt and sugar? How can these kinds of mixtures be separated? Both sand and salt and salt and sugar are too small to be hand sorted. In addition salt and sugar are both the same color! What about a mixture of two clear liquids, such as water and alcohol? What techniques can be used to separate these kinds of mixtures?

Types of Mixtures

The technique used to separate a mixture depends on the chemical and physical properties of the items that are mixed together. Chemical properties are the properties of an atom or molecule that result in chemical reactions. For example, sodium metal will react violently with water, producing hydrogen gas and sodium hydroxide. This is a chemical property of sodium metal.

$$2\ Na\ (s) + 2\ H_2O\ (l) \longrightarrow 2\ NaOH\ (aq) + H_2(g) \quad \text{(see footnote[1])}$$

Pure substances, such as gold or graphite, are separated using chemical methods based on their chemical properties. Physical properties, on the other hand, do not result in chemical reactions, but are properties that make atoms and molecules different without changing them chemically.

In this part of the chapter we will look at ways to separate mixtures using physical properties. Physical properties include color, size, melting point, boiling point, volatility, and solubility. In any mixture where ingredients have different physical properties, these differences can be used to separate the mixtures.

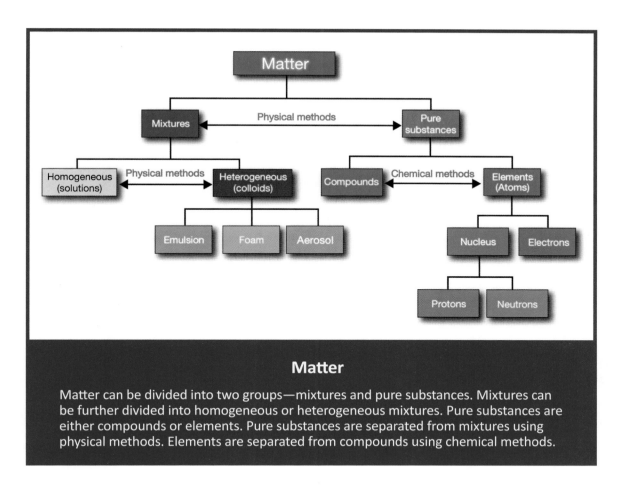

Matter

Matter can be divided into two groups—mixtures and pure substances. Mixtures can be further divided into homogeneous or heterogeneous mixtures. Pure substances are either compounds or elements. Pure substances are separated from mixtures using physical methods. Elements are separated from compounds using chemical methods.

---

[1] The abbreviations next to the chemical formulas give the physical description of the molecules: (s) stands for solid, (l) stands for liquid, (aq) stands for aqueous, and (g) stands for gaseous.

Properties of Mixtures

When deciding how to separate a mixture, it is important to consider all of the physical properties of each component in the mixture. The separation technique used will depend on *differences in the physical properties* between the components of a mixture.

For example, consider a mixture of sand and pebbles. A pebble is much larger than a grain of sand, so there is a significant difference in *size*—a physical property. Both sand and pebbles are made from similar materials (i.e., the materials found in rocks and dirt, such as silicon and quartz). Because they are made of similar materials, they are both solids at room temperature, probably have similar boiling and melting points, and are both insoluble in water. All of these physical properties are similar. The only physical property that is not similar is size. Therefore, a good way to separate sand and pebbles would be to use a technique that separates mixtures based on *physical size*, such as filtering or sieving.

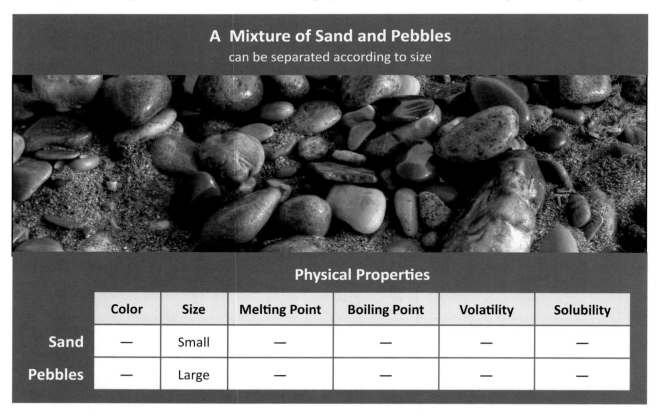

A Mixture of Sand and Pebbles
can be separated according to size

Physical Properties

	Color	Size	Melting Point	Boiling Point	Volatility	Solubility
Sand	—	Small	—	—	—	—
Pebbles	—	Large	—	—	—	—

Now consider a mixture of sand and table salt. The grains of sand and table salt crystals are both similar in size, so they cannot easily be separated using filtering or sieving. However, we know that grains of sand and table salt crystals are not made of the same material. Sand is made mainly of water-insoluble silicon and quartz, but table salt is made of water-soluble sodium chloride. Because sand and table salt have different solubilities in water, they can be separated based on water solubility. We could dissolve the salt in water and pour off the water, separating it from the sand. But this would leave us with a solution of salt in water. How do we get the solid salt back? Water is very volatile but salt is nonvolatile. Volatility refers to a substance's ability to evaporate and become a gas. We can use this difference in volatility by evaporating the water, leaving pure crystalline salt behind. In the end we have pure sand and pure salt.

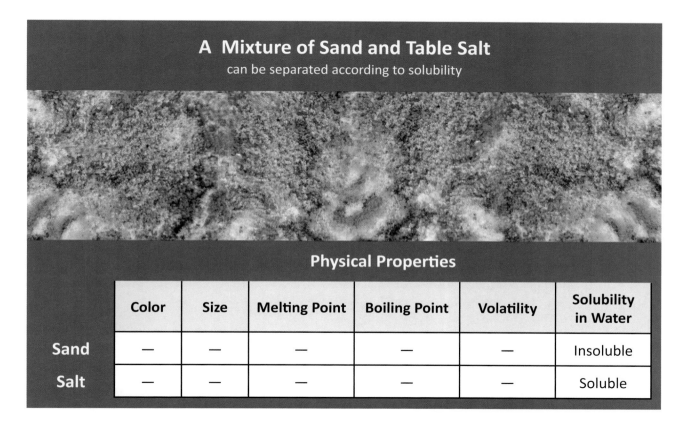

A Mixture of Sand and Table Salt
can be separated according to solubility

Physical Properties

	Color	Size	Melting Point	Boiling Point	Volatility	Solubility in Water
Sand	—	—	—	—	—	Insoluble
Salt	—	—	—	—	—	Soluble

Let's look at a mixture of alcohol and water. A mixture of alcohol and water is much harder to separate than a mixture of sand and pebbles or sand and salt because both water and alcohol are liquids composed of individual molecules. Because they are both on the molecular scale, they cannot be separated based on size. Also, alcohol is soluble in water, so alcohol cannot be separated from water based on water solubility. To find a suitable way to separate alcohol and water we have to look at their other physical properties. For example, if we compare the boiling points for alcohol and water, we see that they are slightly different. Water boils at 100° C and alcohol boils around 80° C. Because alcohol boils at a lower temperature, alcohol is more volatile than water. We can separate a mixture of alcohol and water using their difference in volatility.

A Mixture of Alcohol and Water
can be separated according to solubility

Physical Properties

	Color	Size	Melting Point	Boiling Point	Volatility	Solubility
Alcohol	—	—	—	80°C	higher than water	—
Water	—	—	—	100°C	lower than alcohol	—

9.8 Techniques of Separation

The preceding examples illustrate several circumstances in which techniques of separation based on physical properties can be used. Over time, chemists have developed a number of different methods to separate mixtures based on differences in physical properties. These techniques include filtration, evaporation, distillation, and chromatography.

Filtration

Filtration separates components of a mixture based on the differences in their physical size. A mixture of sand and pebbles can be separated using filtration because their physical sizes are different. To separate a mixture by filtration, a filter is used. A filter can be anything from a metal sheet with large holes in it, to a colander, a sieve, or a piece of paper. The holes in a filter are called pores. The pore size of a filter should be selected so that only part of the mixture will go through the pores with the remaining mixture retained by the filter. The pore size will vary depending on the relative sizes of the components of the mixture. For example, a metal sheet with large holes punched through it may be used to separate large rocks or stones from smaller rocks or sand. On the other hand, a mixture of smaller rocks and fine sand needs to be separated with a filter that has a smaller pore size, such as a sieve or wire mesh.

Chapter 9: Pure Substances and Mixtures 101

In chemistry, paper filters are commonly used to separate chemical precipitates (solids) from the aqueous (water) portion, of a chemical reaction. Paper filters have microscopic pores that allow water to seep through while retaining the solids. For example, the chemical reaction between silver nitrate and sodium chloride produces a water-insoluble precipitate called silver chloride. To separate the silver chloride from the water, a filter paper is used. A simple filtration apparatus consists of filter paper in a funnel placed on top of a collection flask.

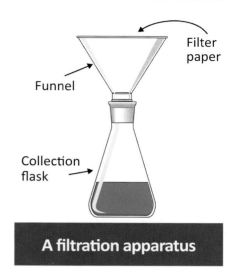

A filtration apparatus

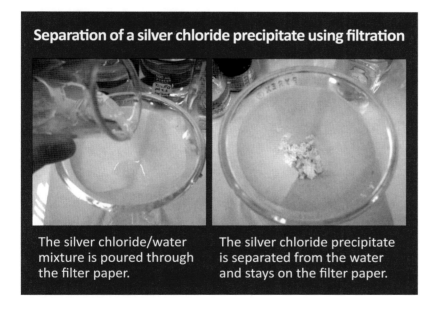

Separation of a silver chloride precipitate using filtration

The silver chloride/water mixture is poured through the filter paper.

The silver chloride precipitate is separated from the water and stays on the filter paper.

The aqueous silver chloride mixture is poured through the filter paper. The water flows through the paper into the collection flask below the funnel, and the silver chloride precipitate is retained on the filter paper.

Evaporation

Evaporation separates components of a mixture based on their differences in *volatility*. Molecules that are volatile have a lower boiling point than molecules that are not volatile. For example, a mixture of water and dissolved salt cannot be separated by filtration because both salt and water molecules are similar in size. But because water is *more volatile* (has a lower boiling point) than salt, a saltwater mixture can be separated using evaporation.

Evaporation is a simple technique that is used in a variety of situations. For example, French chefs use the differences in volatility between alcohol and water to create fine main course dishes or tasty desserts. When alcohol is added to a water mixture of spices or sugars, for example, the mixture is generally heated. Because the alcohol in the mixture is more volatile than the water, it boils off, or evaporates, sooner than water, leaving behind only the added flavoring the dissolved molecules in the alcohol provide.

Distillation

Distillation also separates components of a mixture based on their *differences in volatility*. Distillation is performed using a distillation apparatus (see following illustration). A distillation apparatus is able to capture the more volatile component and cools it back to a liquid, thus separating it from the other components in the mixture. For example, at sea level, water boils at 100° C and ethanol (alcohol) boils at 78.5° C.

A simple distillation apparatus can be used to separate the alcohol from the water. The alcohol–water mixture is heated until the temperature is between 78.5° C and 100° C. Both alcohol and water vapor are formed. But because alcohol is more volatile than water, there is more alcohol vapor than water vapor that goes up the column of the apparatus. As the vapors rise, they cool and recondense onto the walls of the column, forming a new mixture of water and alcohol that has more alcohol than the original mixture.

This new mixture continues to re-evaporate and condense further up the column. Once the vapor mixture reaches the top, it passes through a condenser. The condenser is a long tube that has cold water flowing through it. This water is separated from the alcohol–water vapor mixture by a tube. The condenser is used to cool the vapor mixture back into a liquid. The liquid drips into a collection flask. Most of the water has been separated away from the alcohol-water mixture, leaving the water behind.

Some mixtures, like salt and water or motor oil and gasoline, can be separated nearly completely by distillation because their boiling points are very different. But other mixtures with similar boiling points, like water and alcohol, can only be partially separated.

Chromatography

Chromatography separates components of a mixture using differences in mobility—the difference in how fast each component moves through a given medium. The word chromatography comes from the Greek word *chroma*, which means "color" and *graphe*, which means "to write." Chromatography literally means to "write with color."

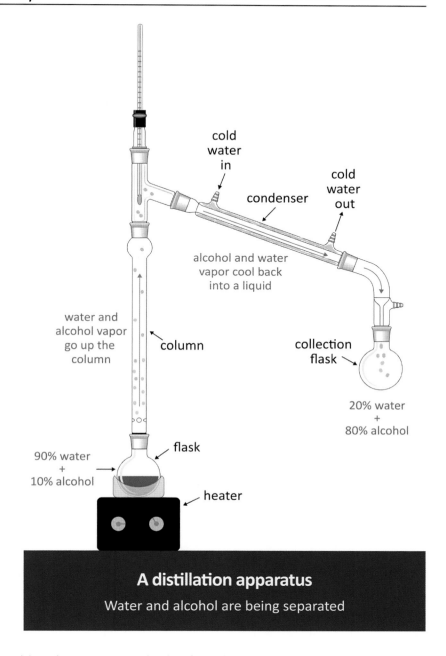

A distillation apparatus
Water and alcohol are being separated

In general, mixtures are separated by chromatography by first dissolving them in a solvent (called the mobile phase) and then passing the dissolved mixture through a finely powdered solid (called the stationary phase). As the mixture in the mobile phase passes over the stationary phase, the components in the mixture will migrate faster or slower (depending on their solubility in the mobile phase) through the stationary phase.

In liquid chromatography, the components of the mixture are dissolved in a liquid. There are two types of liquid chromatography—paper chromatography and column chromatography. Paper chromatography utilizes paper as the stationary phase. In column chromatography the stationary phase is made of silicon beads packed into a column. Liquid chromatography is commonly used to separate larger molecules, such as pigments, proteins, or DNA, which are soluble in a liquid.

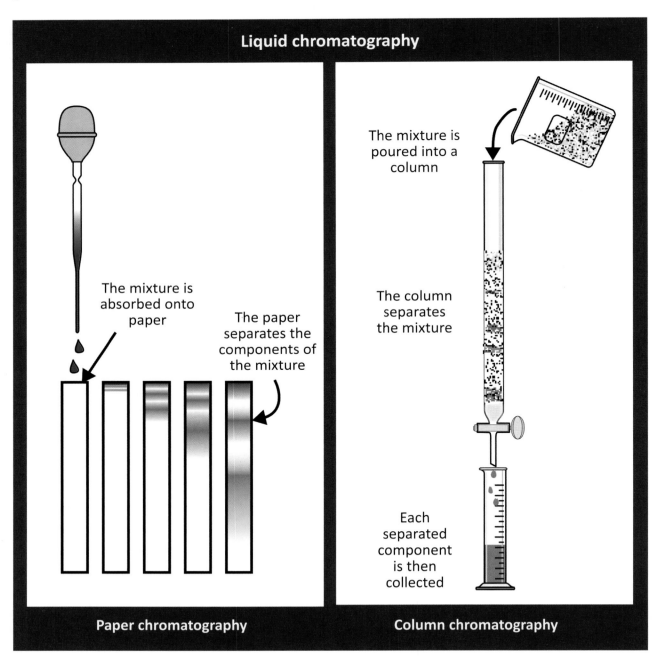

9.9 Summary

- A mixture is made of two or more substances that are physically mixed, but not chemically bonded.

- A homogeneous mixture is a mixture that is the same throughout with molecules that are mixed on a molecular level.

- A heterogeneous mixture is a mixture that is not the same throughout and has particles that are dispersed in clumps or droplets that are usually large enough to be seen.

- Solutions result from one type of molecule (the solute) dissolving into another type of molecule (the solvent).

- Molecules that are "like" each other dissolve.

- The techniques used to separate mixtures depend on the chemical and physical properties of the components in the mixture.

- Separation techniques depend on the differences in physical properties for each component in a mixture, including color, size, melting point, boiling point, volatility, and solubility.

- Filtration separates components of a mixture based on the differences in their physical size.

- Evaporation and distillation separate components of a mixture based on their differences in volatility.

- Chromatography separates components of a mixture based on their differences in mobility through a solid.

9.10 Some Things to Think About

- Explain what you can learn from studying the chart in Section 9.1.

- Explain the difference between an element, a compound, and a mixture.

- Make a list of three substances you think are compounds and another list of three substances you think are mixtures. How did you decide which list to put each item in?

- Describe the difference between a mixture and a compound.

- List three mixtures and three compounds. How did you decide which list to put each item in?

- If you needed to find out if a mixture was a solution or a colloid, what characteristics would you look for?

- Define solute and solvent.

- Explain the rule for solubility: Like dissolves like.

- Explain the difference between hydrophobic and hydrophilic molecules.

- How would you describe the way soap works?

- List some of the physical properties of a mixture of sand and pebbles.

    How would you separate this mixture?

- List some of the physical properties of a mixture of sand and table salt.

    How would you separate this mixture?

- List some of the physical properties of a mixture of alcohol and water.

    How would you separate this mixture?

- Describe a separation technique for each of the following mixtures, and explain why you think this technique will work.

    Water and copper nuggets
    Water and table salt
    Chalk dust and table salt
    Egg whites and water
    Red dye and yellow dye

Chapter 10 Organic Chemistry: The Chemistry of Carbon

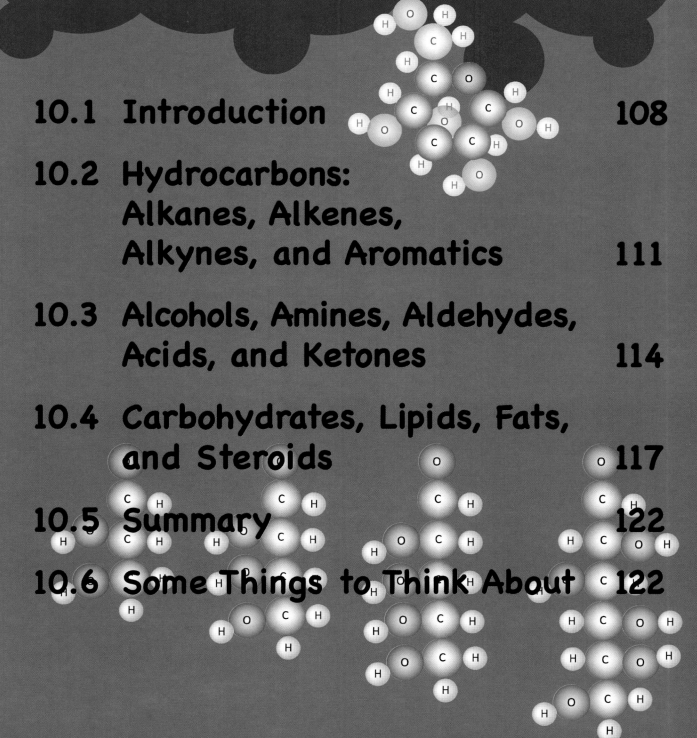

10.1 Introduction   108

10.2 Hydrocarbons:
      Alkanes, Alkenes,
      Alkynes, and Aromatics   111

10.3 Alcohols, Amines, Aldehydes,
      Acids, and Ketones   114

10.4 Carbohydrates, Lipids, Fats,
      and Steroids   117

10.5 Summary   122

10.6 Some Things to Think About   122

10.1 Introduction

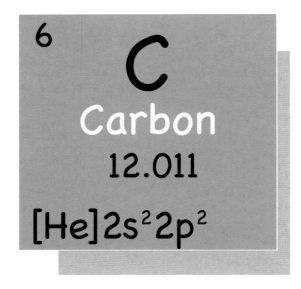

Organic chemistry is a special branch of chemistry that singles out just one element for special consideration—carbon. The chemistry of carbon is especially important because carbon is the most useful and versatile of all the elements in the periodic table. Carbon can be used to form more kinds of molecules than any other element. Carbon also forms some of the strongest chemical bonds known, and it is the only element that can form very large, complex molecules. For these reasons carbon is essential to all of life as we know it. Understanding the chemistry of carbon is important if we are to understand the molecules that fuel not only our bodies but our cars, boats, and airplanes as well.

If we were to make a list of all known molecules (a very, very long list), we would find that the vast majority are based on carbon atoms. Because the chemistry of carbon is so important, the science of chemistry is divided into two main branches: organic chemistry, which deals with carbon-containing molecules, and inorganic chemistry, which deals with everything else. With a few exceptions, all molecules that contain carbon are considered to be organic molecules.[1] Table 10.1 shows some common examples of organic molecules.

There is no limit to the size and complexity of organic molecules. Those shown in Table 10.1 are often called "small molecules" by organic chemists. There are larger organic molecules, such as proteins and DNA found inside living things, that have thousands or even millions of atoms each.

Organic chemistry is a huge subject. At the college level, it is usually taught as a yearlong course all by itself. In this chapter, we will learn about some of the important classes of organic molecules.

---

[1] Carbon dioxide (CO_2), hydrogen cyanide (HCN), graphite, and diamond are not considered by most scientists to be organic substances even though they contain carbon.

Chapter 10: Organic Chemistry: The Chemistry of Carbon

Table 10.1 Common Organic Molecules

Name of Molecule	Chemical Formula	Structure
methane	CH_4	
acetylene	C_2H_2	
ethanol	CH_3CH_2OH	
chloroform	$CHCl_3$	
acetic acid	CH_3COOH	
formaldehyde	H_2CO	
glycine	H_2NCH_2COOH	
benzene	C_6H_6	
octane	C_8H_{18}	

Classes of Organic Molecules

Because there are so many different kinds of organic molecules, it is useful to classify them into groups, or classes, and learn about the groups one at a time. The most common groups of organic molecules are alkanes, alkenes, alkynes, aromatics, alcohols, amines, aldehydes, acids, and ketones. Table 10.2 shows examples of each of these kinds of organic molecules.

Table 10.2 General Groups (Classes) of Organic Molecules

Name	Description	Structure	
alkanes	Molecules that contain only carbon and hydrogen and only single bonds between carbon atoms	(structure)	methane
alkenes	Molecules with one or more double bonds between two carbon atoms	(structure)	ethene
alkynes	Molecules with one or more triple bonds between two carbon atoms	H−C≡C−H	ethyne
aromatics	Molecules containing a benzene ring	(structure)	bromobenzene
alcohols	Molecules with an -OH attached to a carbon atom	(structure)	methanol (methyl alcohol)
amines	Molecules with a $-NH_2$ attached to a carbon atom	$CH_3-C(CH_3)(CH_3)-NH_2$	butylamine
aldehydes	Molecules with −C(=O)−H	$CH_3-C(=O)-H$	ethanal (acetaldehyde)
acids	Molecules with −C(=O)OH	$CH_3-C(=O)-OH$	ethanoic acid (acetic acid)
ketones	Molecules with >C−C(=O)−C<	$CH_3-C(=O)-CH_3$	2-propanone (acetone)

Chapter 10: Organic Chemistry: The Chemistry of Carbon

There are also many other types of organic molecules, but these are sufficient to understand the basics of organic chemistry and most of the chemistry of biology.

10.2 Hydrocarbons

The first four kinds of organic molecules (alkanes, alkenes, alkynes, and aromatics) are all hydrocarbons—that is, they contain only hydrogen and carbon. They are all very nonpolar, flammable, and similar in both appearance and touch (for the solids and liquids).

Alkanes

The simplest organic molecules are the alkanes which have only single bonds and contain only carbon and hydrogen. Table 10.3 shows some common examples of alkanes.

Table 10.3 Alkanes

Name	Structure	Properties	Uses
methane	CH_4	colorless odorless gas	main component of natural gas
ethane	CH_3CH_3	colorless odorless gas	a component of natural gas
propane	$CH_3CH_2CH_3$	gas	can be liquefied at high pressure (LP gas used for camp stoves and gas grills)
n-butane	$CH_3CH_2CH_2CH_3$	gas	can be liquefied at low pressure (used in butane lighters)
isobutane	$CH_3CH(CH_3)CH_3$	gas	can be liquefied at low pressure
n-pentane	$CH_3CH_2CH_2CH_2CH_3$	liquid	gasoline-like fuel
n-decane	$CH_3(CH_2)_8CH_3$	liquid	a bit oily
n-eicosane	$CH_3(CH_2)_{18}CH_3$	waxy solid	found in paraffin waxes in candles
polyethylene	$CH_3(CH2)_nCH_3$	common plastic	(milk bottles, etc.)

An alkane molecule can be any size. The shortest is methane, CH_4, with only one carbon atom. Ethane has two carbons, propane has three carbons, and so on up to eicosane with 20 carbons, and polyethylene, which may have hundreds or even thousands of carbons.

The small alkanes are gases, the medium ones (from pentane on) are liquids, and the larger ones are solids. All the alkanes are very nonpolar. The liquids are gasoline-like or oily and act as solvents for nonpolar substances. The solids are waxes (like paraffin) or plastics, with a waxy, greasy feeling to the touch, and in fact, this is the way all nonpolar substances feel. Alkanes all burn in air and are often used as fuels (natural gas, LP gas, butane lighters, gasoline, candles, etc.).

Alkenes and Alkynes

An alkene is any organic molecule with a carbon-to-carbon double bond, and an alkyne is any molecule with a carbon-to-carbon triple bond. Tables 10.4A and 10.4B show a few common examples (ethene, butene, ethyne, etc.). Like the alkanes, the smaller alkenes and alkynes are gases, the medium ones are nonpolar liquids, and the larger ones are waxy solids or plastics. Also like the alkanes, the alkenes and alkynes burn in air. Gasoline is a mixture of many organic molecules, including large amounts of both alkanes and alkenes. Acetylene, the smallest alkyne, burns so hot that it is used in welding and cutting torches.

Table 10.4A Alkenes

Name	Structure	Uses
ethene (ethylene)	$H_2C=CH_2$	plant hormone that causes ripening of fruit
propene (propylene)	$CH_3-CH=CH_2$	monomer used to make polypropylene, a common polymer
1-butene (butylene)	$H_2C=CH-CH(H)-CH_3$	monomer used to make polybutylene, a common polymer
2-butene	$CH_3-CH=CH-CH_3$	used in the production of gasoline

Table 10.4B Alkynes

Name	Structure	Uses
ethyne (acetylene)	H—C≡C—H	used in welding and cutting torches
propyne	CH_3—C≡C—H	used in welding torches
1-butyne	H—C≡C—CH(H)—CH_3	used in the synthesis of organic compounds
2-butyne	CH_3—C≡C—CH_3	used in the synthesis of organic compounds

Aromatics

The last and most complex of the hydrocarbons are the aromatic molecules. (They usually smell good!) There are many different kinds of aromatic molecules, but we will focus on the simplest one, benzene.

As you can see in Figure 10.1, benzene is a ring of six carbon atoms and six hydrogens in the shape of a hexagon. There are three double bonds alternating with three single bonds around the ring.

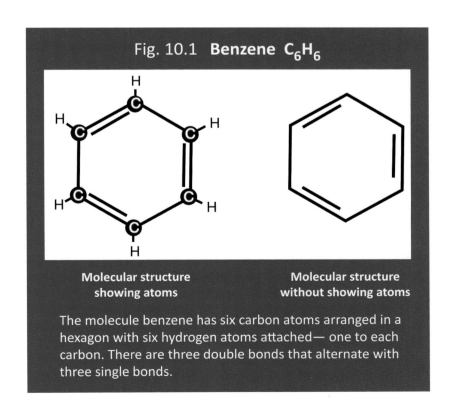

Fig. 10.1 Benzene C_6H_6

Molecular structure showing atoms

Molecular structure without showing atoms

The molecule benzene has six carbon atoms arranged in a hexagon with six hydrogen atoms attached— one to each carbon. There are three double bonds that alternate with three single bonds.

10.3 Alcohols, Amines, Aldehydes, Acids, and Ketones

Alcohols and Amines

The simplest organic molecules beyond the hydrocarbons are the alcohols and amines as shown in Table 10.5. An alcohol is any molecule with a -OH group attached to a carbon atom, and an amine is any molecule with a $-NH_2$ group attached to a carbon atom. Both the -OH group and the $-NH_2$ group are very polar, so alcohols and amines are usually polar as well. They tend to dissolve well in water, and the liquid alcohols, especially, can act as solvents for other polar molecules.

Table 10.5 Some Simple Alcohols and Amines

Name	Structure	Uses
methanol	CH_3OH	methyl alcohol—wood alcohol
ethanol	CH_3CH_2OH	ethyl alcohol—"alcohol" in wine, beer, etc.
1-propanol	$CH_3CH_2CH_2OH$	used as a solvent
2-propanol	$CH_3\underset{\underset{\displaystyle OH}{\mid}}{C}H_2CHCH_3$	isopropyl alcohol—"rubbing alcohol"
1-butanol	$CH_3CH_2CH_2CH_2OH$	a solvent
ethylene glycol	$HOCH_2CH_2OH$	antifreeze
glycerol	$HOCH_2\underset{\underset{\displaystyle OH}{\mid}}{C}HCH_2OH$	glycerine
methylamine	CH_3NH_2	used in producing agricultural chemicals
ethylamine	$CH_3CH_2NH_2$	used in synthesizing organic molecules
1-propylamine	$CH_3CH_2CH_2NH_2$	used in chemical analysis and synthesis

Both alcohols and amines are very important for living things. All the carbohydrates, including sugars, starches, and cellulose, are large alcohols with many -OH (alcohol) groups each. The amino acids from which all proteins are built are amines and acids.

Aldehydes, Acids, and Ketones

In our next group, all the molecules have a characteristic feature called a carbonyl, which is a carbon atom that is double bonded to an oxygen [>C=O]. Like alcohols and amines, the carbonyl is polar, so all the molecules in this section are also polar, though not as much so as the alcohols and amines. Depending on what is next to the carbonyl group, the molecule may be an aldehyde, an acid, or a ketone:

- An aldehyde is any molecule that: Has only an H atom on one side of the carbonyl.
- An acid is any molecule that: Has an -OH group next to the carbonyl.
- A ketone is a molecule that: Has carbon atoms on both sides.

Table 10.6 on the following page shows simple examples of all three types.

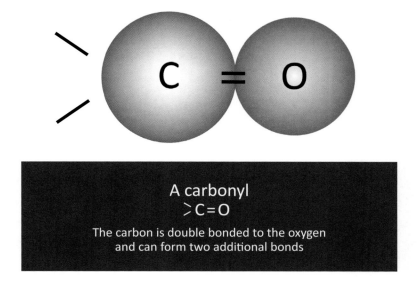

A carbonyl
>C=O

The carbon is double bonded to the oxygen
and can form two additional bonds

Table 10.6 Aldehydes, Acids, and Ketones

Name	Structure	Uses
methanal	H-C(=O)-H	formaldehyde, preservative
ethanal	$\text{CH}_3\text{-C(=O)-H}$	used in making perfumes and food flavors
octanal	$\text{CH}_3\text{-CH}_2\text{CH}_2\text{CH}_2\text{CH}_2\text{CH}_2\text{CH}_2\text{C(=O)-OH}$	citrus aroma found in citrus oils
formic acid	H-C(=O)-OH	both an acid and an aldehyde; used in food preservatives
ethanoic acid (acetic acid)	$\text{CH}_3\text{-C(=O)-OH}$	vinegar
glycine	$\text{NH}_2\text{-CH}_2\text{-C(=O)-OH}$	the simplest amino acid (both an acid and an amine); used to make proteins
alanine	$\text{NH}_2\text{-CH(CH}_3\text{)-C(=O)-OH}$	the next simplest amino acid used to make proteins
benzoic acid	C₆H₅-C(=O)-OH (benzene ring with carboxylic acid)	both an aromatic and an acid often used as a preservative in sodas (under the name sodium benzoate—look on the label on a soda)
propanone (acetone)	$\text{CH}_3\text{-C(=O)-CH}_3$	a solvent
butanone	$\text{CH}_3\text{CH}_2\text{-C(=O)-CH}_3$	methyl ethyl ketone or MEK, a common solvent
2-pentanone	$\text{CH}_3\text{CH}_2\text{CH}_2\text{-C(=O)-CH}_3$	food flavoring and perfumes

10.4 Carbohydrates, Lipids, Fats, and Steroids

Carbohydrates

Carbohydrates are molecules that are essential for living things. Carbohydrates are found as small simple sugars and large complex polymers. Small simple sugars are called monosaccharides.

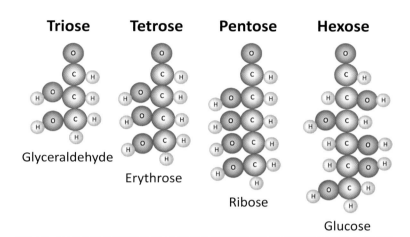

Fig. 10.2 **Simple sugar carbohydrates**

Glyceraldehyde has 3 carbons, erythrose has 4 carbons, ribose has 5 carbons, and glucose has 6 carbons.

The smallest monosaccharides have three carbon atoms. These are called trioses. Larger simple sugars with four, five, six, and seven carbons are called tetroses, pentoses, hexoses, and heptoses, respectively. Glyceraldehyde, the simplest sugar, is a triose with only three carbons. Erythrose is a tetrose with four carbon atoms. Ribose is a pentose and has five carbons. And glucose is a hexose and has six carbon atoms.

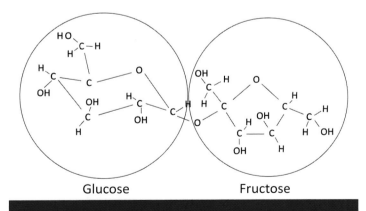

Fig. 10.3 **Sucrose** (table sugar)
a disaccharide of glucose and fructose

When single sugars are added one to another, larger and more complex carbohydrates are formed. When two monosaccharides are connected, the molecule becomes a disaccharide (di- means "two"). Sucrose, or table sugar, is a disaccharide of a single glucose and a single fructose.

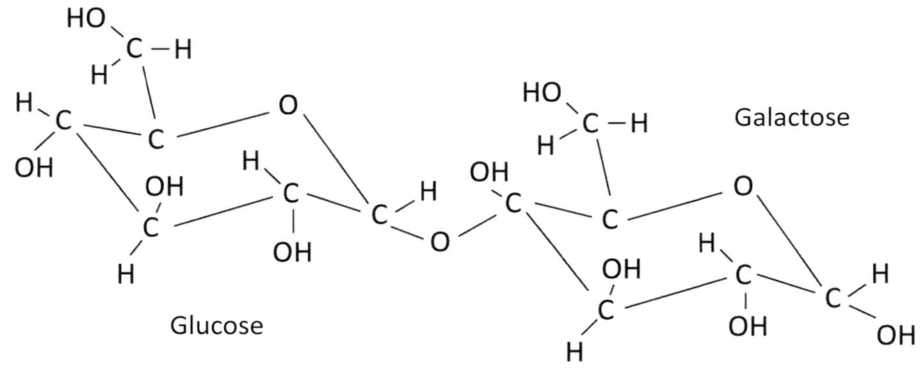

Fig. 10.4 Lactose (milk sugar)
a disaccharide of glucose and galactose

Lactose, the sugar found in milk, is made of a glucose and a galactose.

When a few (more than two) saccharides are added together, the molecule is called an oligosaccharide (oligo- means "few"), and when many saccharides are added together in a long chain, the molecule is called a polysaccharide [poly- means "many"].

There are two major types of polysaccharides: structural polysaccharides and storage polysaccharides. As the name implies, structural polysaccharides are carbohydrates that are primarily involved in plant cell walls and insect exoskeletons. Storage polysaccharides, on the other hand, are used for storing food energy.

Cellulose and chitin are two structural polysaccharides. Cellulose is the primary structural polysaccharide for cell walls in plants.

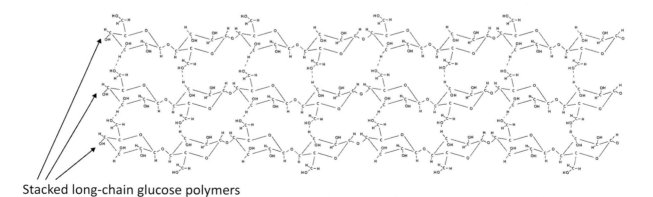

Stacked long-chain glucose polymers

Fig. 10.5 Cellulose
Cellulose is made up of long-chain polymers that stack on top of each other.

Chitin is the main structural polysaccharide found in the exoskeletons of insects, spiders, and crustaceans, and it forms similar stacked layers.

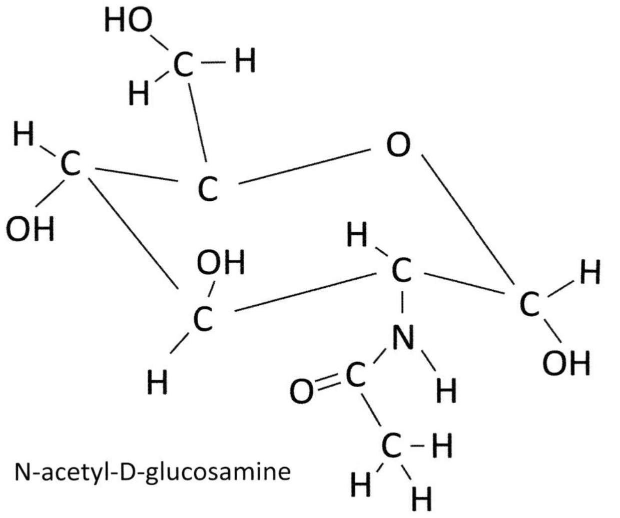

N-acetyl-D-glucosamine

Fig. 10.6 The monomer in chitin

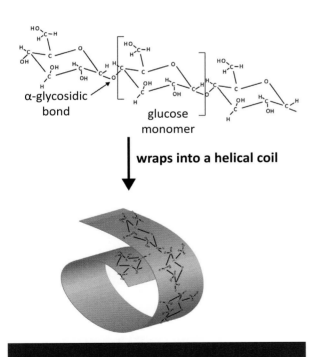

Fig. 10.7 Amylose
Amylose wraps into a helical coil.

Starch and glycogen are the primary storage polysaccharides found in plants and animals. Starches are found only in plants, and glycogen is the storage polysaccharide found in animals.

Starch is composed of two different polysaccharides: amylose and amylopectin. Amylose is a linear polymer of glucose monomers linked together. A monomer is a particular group of atoms that repeats within a molecule. (See Chapter 11.)

Amylose forms a spiral coil, or helix which looks similar to a Slinky®. It is this unique structure that binds iodine molecules, creating a deep purple color in iodine-stained foods.

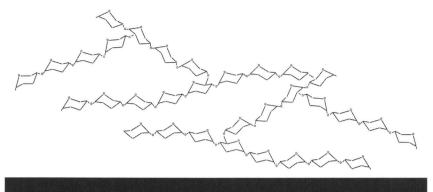

Fig. 10.8 **Amylopectin**
Amylopectic is a branched polymer.

Amylopectin is also a polymer of linked glucose monomers. But unlike amylose, amylopectin is branched instead of linear and does not form a helical coil.

Glycogen is composed of glucose molecules similar to amylopectin but with many more branches. Glycogen is present in nearly all cells, but in humans it is found primarily in liver and skeletal muscle cells.

Lipids: Fats and Steroids

Another important group of nutrients required for the healthy maintenance and function of our bodies are the lipids. Lipids include fats, sterols, waxes, fat-soluble vitamins, and other molecules. Fats allow the body to absorb fat-soluble vitamins, provide energy, and are an essential component of cellular membranes.

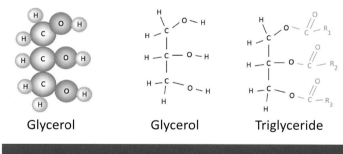

Fig. 10.9 **Glycerol**
Glycerol is a small three-carbon carbohydrate that is used to make triglyceride.

The most common fats in living things are made from glycerol. Glycerol is a small three-carbon carbohydrate. Fats are made of a derivative of glycerol, called a triglyceride.

Have you ever wondered why at room temperature animal fat is solid but vegetable oil is liquid? Both animal fat and vegetable oil are fats made of a triglyceride and three long chains of hydrocarbons, but an animal fat has no double bonds in its hydrocarbons, whereas a vegetable oil does. If there are no double bonds, the fat is called saturated. If the fat does contain double bonds, it is called unsaturated. Saturated fats have a higher melting temperature than unsaturated fats. Because animal fats are typically saturated and have no double bonds, they are solids at room temperature. Vegetable fats are unsaturated, do have double bonds, and have a lower melting temperature than animal fats, making them liquid at room temperature.

Finally, another set of important nutrients is called steroids. Steroids are found in both plants and animals and are among the most important natural products. Steroids are involved in sex hormones, bile acids, and the formation of animal membranes.

Cholesterol is the most common steroid found in animals. Cholesterol is a type of lipid found in the brain and spinal column tissues of humans and is the major component in the plasma membranes of animal cells.

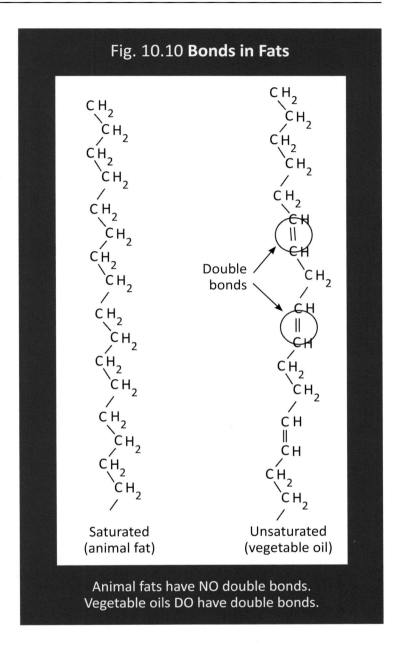

Cholesterol is the primary starting material of steroid hormones. Although there has been significant dietary controversy over cholesterol, it is an important biochemical molecule and a vital nutrient for the proper health and maintenance of our bodies. (See Fig. 10.11)

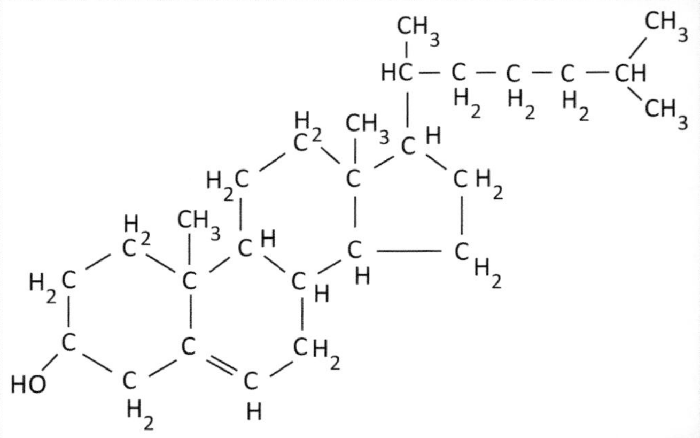

Fig. 10.11 Cholesterol

Cholesterol is a steroid and is an important molecule for our bodies.

10.5 Summary

- Organic chemistry deals with carbon-containing compounds.

- Alkenes, alkynes, alkanes, and aromatics are groups of organic molecules that contain only hydrogen and carbon.

- Carbohydrates are made of simple sugars or chains of simple sugars and provide energy for living things.

- Cellulose and chitin are structural polysaccharides. Cellulose is found in plants, and chitin is found in the exoskeletons of insects and crustaceans. Starch and glycogen are storage polysaccharides found in both plants and animals.

- An unsaturated fat has double bonds. A saturated fat has no double bonds.

10.6 Some Things to Think About

- How would you describe the difference between organic and inorganic chemistry?

- What are the main characteristics of alkanes, alkenes, alkynes, and aromatics as a group?

- How would you describe a carbonyl?

- What is one reason that esters are important?

- What is the difference between saturated and unsaturated fats?

Chapter 11 Polymers

11.1 Introduction	124
11.2 Polymer Structure	124
11.3 Polymer Properties and Reactions	127
11.4 Summary	134
11.5 Some Things to Think About	134

11.1 Introduction

Molecules with repeating units are called polymers. Poly means "many." The word root mer comes from the Greek word *meros*, which means "unit," so a polymer is a molecule of "many units." Polymers are found everywhere. Both naturally occurring polymers, such as polysaccharides, and man-made polymers, such as plastics and Styrofoam, are found in every facet of life.

Polymers serve as the primary material for clothing, boats, footballs, roof tiling, garbage bags, and even spacecraft. Before the early 1900s only naturally occurring polymers were available. Structural items for housing or carriages were made of wood, and clothing was primarily made of cotton, wool, flax, or silk fibers. Synthetic polymers, in contrast, are man-made molecules that have been developed by organic chemists. The first synthetic polymer, Bakelite, was produced in 1909, followed by rayon in 1911. Today, several hundred synthetic polymers are used in everyday life. Nylon, Styrofoam, vinyl, and all plastics are synthetic polymers used for everything from clothing to bottles, paints, and pipes. In this chapter, we focus on synthetic polymers. Biological polymers, such as proteins and DNA, are the focus of Chapter 12.

11.2 Polymer Structure

Monomers

It was once believed that polymers were simply colloidal aggregates of smaller molecules, but in the 1950s polymers were discovered to be composed of a sequence of repeating units linked together by covalent bonds. The individual units of a polymer are called monomers.

Monomers can be anything from simple double-bonded hydrocarbons, called alkenes, to more complex molecules containing ring structures. When the repeating monomer units are all identical, the polymer is called a homopolymer. For example, polyethylene is a homopolymer made from a simple alkene, ethylene (also called ethene). The double bond in ethene is the functional group that allows the monomers to be hooked together. Functional groups are special sites on larger molecules at which chemical reactions can take place and that can be used by organic chemists to build new molecules. Polyethylene, polyamide (nylon), polystyrene (Styrofoam) and polyvinyl chloride (PVC) are all homopolymers made of a single repeating monomer unit.

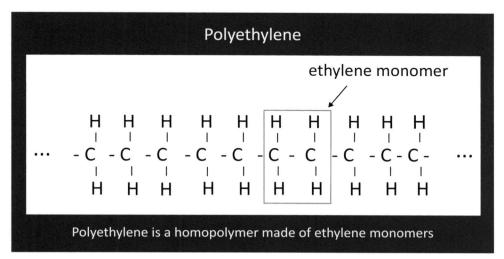

Polyethylene is a homopolymer made of ethylene monomers

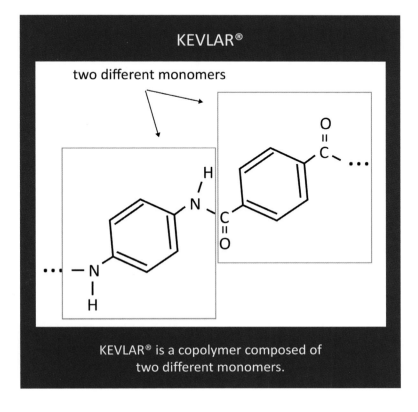

KEVLAR® is a copolymer composed of two different monomers.

If more than one type of monomer is used, the polymer is called a copolymer. For example, KEVLAR®—a very strong polymer used for bulletproof vests—is made from two complex monomers: one with two acid functional groups and one with two amino groups.

Copolymers can have random repeats or nonrandom repeats with a variety of different combinations. Most polymers are neither completely random nor non-random, but a mixture of both.

Linear or Branched

Polymers can be linear or branched. In a linear polymer every monomer unit is connected to the next monomer unit, one after another, end to end. Polyethylene is an example of a linear polymer. Linear polymers vary in length and can have several thousand monomer units linked together. Linear polymers, however, do not stay as a stretched out chain of monomer units, but instead roll up into a random coil. As the name implies, a random coil is a random

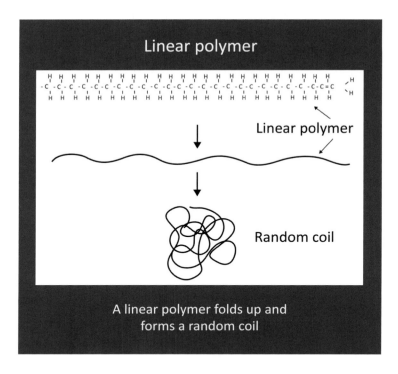

A linear polymer folds up and forms a random coil

folding of the long-chain polymer into a more compact ball. The degree to which a polymer will coil depends on several factors, including the type of monomer units in the polymer and the number of single and double bonds covalently linking the monomer units together.

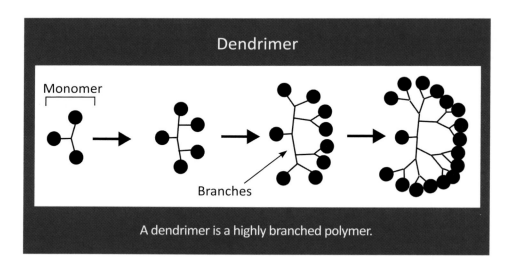

A dendrimer is a highly branched polymer.

Polymers can also be branched, with chains that are connected to each other in the middle. They can be slightly branched, with only a few branches coming off of the polymer backbone, or highly branched, with many branches connecting additional chains of monomers to the polymer backbone. In fact, polymers can be so extensively branched that there is no backbone at all. Dendrimers are extensively branched polymers that grow with a constantly increasing number of branches. Dendrimers are used as films and fibers because they have excellent surface properties.

Cross-links

Polymer molecules can also be connected to each other through cross-links. A cross-link is a covalent bond between any two polymer chains. Cross-links can form as the polymer is built up, or polymerized, or they can be added later by additional chemical reactions. (See Section 11.3.) Cross-linking occurs primarily through double bonds or functional groups on the monomers, such as alcohols (-OH groups attached to carbon), carbonyl groups, and C=O.

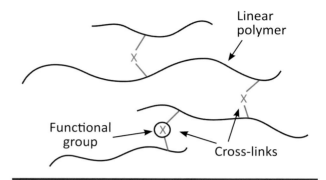

Cross-links connect linear polymer chains together.

11.3 Polymer Properties and Reactions

Polymer Structure

The physical properties of a polymer (its hardness, stretchiness, melting temperature, etc.) are largely determined by the structure of the polymer and the way in which the polymer chains pack with each other in a solid. As we have already seen, polymers can be made of a single type of monomer or composed of two or more monomers, forming a copolymer. We also saw how polymers can be linear, branched, or cross-linked. These characteristics help determine the physical properties of the polymer (i.e., the melting or boiling point, the viscosity, the hardness or softness, and whether or not the polymer is brittle or elastic).

Thermoplastics are the category of polymers people think of when they hear the word *plastic*. Polyethylene and polystyrene are thermoplastics. Thermoplastics are hard at room temperature but soften when heated. Because thermoplastics soften when heated, they can be easily molded into a variety of shapes and structures. Football helmets, computer keyboards, and hula hoops are made of thermoplastics.

Because synthetic fibers (those that are man-made) such as nylon or Dacron® have different monomers from those of thermoplastics, they also have different structural properties. Nylon and Dacron® can be drawn out into long, thin fibers that can be used to make thread which can then be woven into cloth. Polymer fibers of nylon and Dacron® can be quite strong because of their long polymer chains.

Elastomers are polymers that have the ability to stretch and spring back to their original shape. Natural rubber is an elastomer.

Polymer Addition Reactions

Polymers are built by hooking monomers together in chemical reactions. This can be done using many different reactions, but we will consider only two: addition reactions and condensation reactions.

Addition reactions link together molecules using double bonds as the functional group. The simplest addition polymer is polyethylene, made by hooking together molecules of ethene (which is commonly called ethylene).

The addition reaction starts with the formation of a free radical. A free radical is just a "dangling bond"—an unbonded electron that is very reactive. Chemists often put special free radical initiators into their reactions to create free radicals and get the reaction started. In the following example for polyethylene, when the free radical on one molecule encounters the double bond on another ethene molecule, it "steals" one of the two electrons of the ethene. This causes a new carbon–carbon bond to be formed, linking the two molecules together. But one of the electrons is still left over, so it becomes a new free radical.

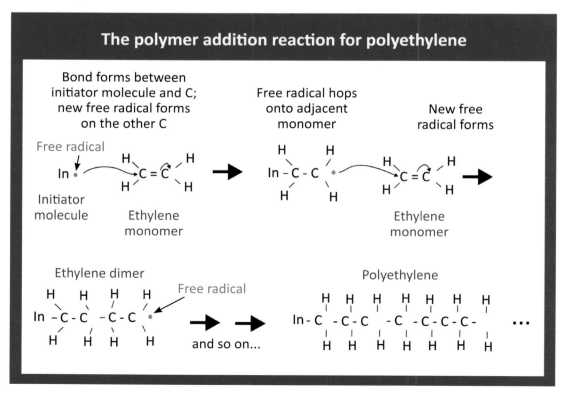

There is now a new, bigger molecule with four carbons, but it still has a reactive free radical on one end. Now another ethene can encounter the free radical, bond to the growing chain, and form a six-carbon molecule with a free radical on *its* end. In this way, we can go on adding new units to the chain, two carbons at a time, until it is thousands of carbons long.

Termination, Branching, Cross-linking

In real life, no chemical reaction happens perfectly. Side reactions may destroy the active free radical and thus stop the polymer chain from growing. This is called chain termination. Also, in the reaction vessel, many polymer chains can get started and grow at the same time. If the free radical on one growing chain hits the middle of a second chain, the polymers can become branched or cross-linked.

By controlling temperature and pressure and by using certain chemical additives, chemists can control how these side reactions happen to some extent. This allows us to make many different forms of polyethylene: some with short chains, some with long chains, some with no branches, some with many branches, some with no cross-links, and some with many cross-links.

The size, branching, and cross-linking of a polymer greatly affect its properties, as we saw earlier. By adjusting these properties, the chemists who synthesize polymers can make polyethylene hard or soft, rubbery or brittle, easily melted or hard to melt.

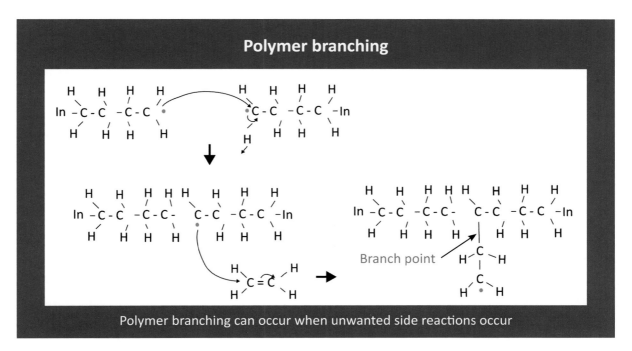

Polymer branching can occur when unwanted side reactions occur

Polyethylene is All Around You

Polyethylene is one of the most common plastics in use today. If you look at the bottom of any plastic bottle, you will likely find a little triangle with the letters "HDPE" or "LDPE." These letters stand for "high density polyethylene" or "low density polyethylene." These are just two types of the many forms of polyethylene. High density polyethylene is made from long, unbranched chains that pack well together. This makes HDPE quite hard and tough. It is used for bottles and other items that need to be strong

or are meant to be used for a long time. Low density polyethylene is made from branched polyethylene chains that do not pack together as well. Thus, it is softer and more easily degraded. It is used for disposable items, like lunch bags, that do not need to be very strong and are meant to be used only once.

Other Addition Polymers

Polyethylene is far from the only polymer made using addition reactions. The following table shows several other monomer molecules with double bonds that can be chained together to form polymers.

Most of these monomers form polymers with side chains—that is, with something that dangles off the main backbone of the polymer. Side chains greatly affect the properties of the polymer. For example, polyvinyl alcohol has -OH side chains. This means that unlike polyethylene, which is extremely nonpolar and insoluble, polyvinyl alcohol is very polar and dissolves easily in water. It can even be eaten, so it is used as a thickener in foods!

Common Polymers and Their Uses

Monomer Name	Structure (monomer in red)	Polymer Name	Common Uses
Ethene (ethylene)	-CH₂-CH₂-CH₂-CH₂-CH₂-CH₂-	polyethylene	bags, films, toys, computer keyboards
Vinyl chloride	-CH₂-CHCl-CH₂-CHCl-CH₂-CHCl-	polyvinyl chloride	PVC pipes, raincoats
Vinyl alcohol	-CH₂-CHOH-CH₂-CHOH-CH₂-CHOH-	polyvinyl alcohol	coatings, thickener in glues, "slime"
Styrene	-CH₂-CH(C₆H₅)-CH₂-CH(C₆H₅)-CH₂-CH(C₆H₅)-	polystyrene	foam insulation, drinking cups
Methyl methacrylate	-CH₂-C(CH₃)(COOCH₃)-CH₂-C(CH₃)(COOCH₃)-CH₂-C(CH₃)(COOCH₃)-	polymethyl methacrylate (PMMA)	Plexiglas™, Lucite®, clear plastic sheets
Tetrafluoroethylene	-CF₂-CF₂-CF₂-CF₂-CF₂-CF₂-	polytetrafluoroethylene (PTFE) Teflon®	nonstick coatings, watertight seals

Forming Polymers by Condensation Reactions

In Chapter 10, we learned that esters can be made by hooking together an acid and an alcohol:

$$CH_3\text{-}\overset{\overset{O}{\|}}{C}\text{-}OH + CH_3\text{-}CH_2\text{-}OH \longrightarrow CH_3\text{-}\overset{\overset{O}{\|}}{C}\text{-}O\text{-}CH_2CH_3 + H_2O$$

acetic acid   ethanol   ethyl acetate (an ester)

Similarly, we can form amides by linking an *acid* and an *amine*:

$$R-\overset{\overset{O}{\|}}{C}-OH + H_2NR' \longrightarrow R-\overset{\overset{O}{\|}}{C}-\underset{\underset{H}{|}}{N}-R' + H_2O$$

organic acid amine amide

Both of these reactions are condensation reactions. A condensation reaction is a chemical reaction in which two monomers combine to form a new molecule, giving off a by-product such as water. This type of reaction is very useful for chaining monomer units together into long-chain polymers called condensation polymers.

One of the first and still one of the most important of the condensation polymers is nylon-6,6. Nylon is a polyamide, which means it is linked together with amide linkages. It is made with two monomers–one is a double acid, and the other is a double amine. One of the acid groups and one of the amine groups can react to form an amide linkage.

Now the molecule has two monomers but still has an acid at one end and an amide at the other, so it can go on adding new monomers. Each additional monomer makes the chain six carbons and one nitrogen longer. It is possible to grow very long chains in this way.

$$OH-\overset{\overset{O}{\|}}{C}-(CH_2)_4-\overset{\overset{O}{\|}}{C}-OH \quad + \quad H_2N-(CH_2)_6-NH_2 \longrightarrow$$

adipic acid   hexamethylenediamine

$$-\overset{\overset{O}{\|}}{C}-(CH_2)_4-\overset{\overset{O}{\|}}{C}-\underset{\underset{H}{|}}{N}-(CH_2)_6-\underset{\underset{H}{|}}{N}-\overset{\overset{O}{\|}}{C}-(CH_2)_4-\overset{\overset{O}{\|}}{C}-\underset{\underset{H}{|}}{N}-(CH_2)_6-\underset{\underset{H}{|}}{N}-$$

Nylon-6,6

The formation of nylon-6,6 is a condensation reaction between two monomers.

Another common class of condensation polymers is the polyesters, formed by hooking monomers together with ester linkages. For example, a very widely used polyester, known by the trade name Dacron®, is made from a double acid monomer and a double alcohol monomer. Similar to nylon-6,6, the new two-monomer molecule forming the polyester still has functional groups at both ends, and so can keep on growing:

OH- CH₂ - CH₂- OH + OH - C(=O) - ⬡ - C(=O) - OH ⟶

ethylene glycol   terephthalic acid

- O - C(=O) - ⬡ - C(=O) - O - CH₂ - CH₂ - O - C(=O) - ⬡ - C(=O) - O - CH₂ - CH₂ - O -

Polyester

The formation of Dacron® is a condensation reaction between two monomers.

The monomers for condensation reactions may have two alcohols, or two amines, in which case they must be paired with a monomer that has two acids. Or a monomer may have one acid and one alcohol so that ester linkages can be formed "tip to tail." In this case, only one kind of monomer is needed. Alternatively, a monomer may have one acid and one amine, so that amide linkages can be formed "tip to tail." The most important example of this case is the amino acids.

When amino acids form a long chain of amide linkages, the resulting polymer is called a protein. As we will see in the next chapter, proteins are the main molecular machines that make all of life possible.

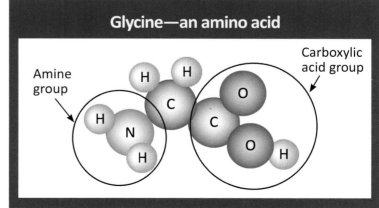

Glycine—an amino acid

As we have seen, chemists have created new synthetic polymers, such as thermoplastics, that can be molded into many different objects and polymers that can be pulled out into fibers to use in making clothing. These synthetic polymers have made it possible for us to improve the overall quality of life for many people. Without synthetic polymers, we would not have been able to manufacture the devices needed to cure certain medical conditions, travel at the speed of sound, or fly to the moon.

11.4 Summary

- A polymer is a molecule of "many units" each of which is a monomer.

- Polymers can contain one type of monomer, or can be composed of two or more different monomer units.

- Polymers can be linear or branched or contain cross-links.

- Polymers, such as polyethylene, can be formed by addition reactions, or they can be formed by condensation reactions, as is Dacron®.

- A condensation monomer can have two different reactive groups, such as amino acids, that allow "tip-to-tail" polymer formation.

11.5 Some Things to Think About

- Why do you think polymers are important to chemists?

- Explain what a monomer is.

- Describe some different polymer structures.

- Describe an addition reaction.
  Describe a condensation reaction.
  How are they different?

Chapter 12 Biological Polymers

12.1 Introduction   136

12.2 Amino Acid Polymers: Proteins 140

12.3 Nucleic Acid Polymers   148

12.4 RNA   152

12.5 Building Biological Machines   153

12.6 Summary   156

12.7 Some Things to Think About   158

12.1 Introduction

As we discovered in the last chapter, chains of repeating units are called polymers. We found that polymers have different properties determined by the type of monomer unit(s) that make up the polymer and by how the polymer chain interacts with itself or other polymer chains in the form of cross-links. We found that polymers can be synthetic (made by humans) or occur in living things. We will now look at two very important biological polymers found in living things. These are amino acid polymers and nucleic acid polymers.

Polymers in Living Things

We now know that living things are composed of a system of highly organized and specialized molecules. Vitamins, fats, carbohydrates, and steroids are all part of the intricate system that keeps our bodies healthy. There are many types of biological polymers that are important for the overall function of biological cells. Two such polymers are amino acid polymers, called proteins, and nucleic acid polymers, which are DNA and RNA. These two types of polymers are specifically designed to carry out a number of different and very important functions inside the cell. These polymers are found in all living things, and without them, living things would not exist.

Living Cells Are Like Tiny Computers

It is not much of an exaggeration to say that all living things are made of biological polymers, especially proteins and DNA. Although there are many molecules inside living things, proteins, DNA, and RNA are the main machinery and information molecules. We can use a computer as a simple analogy to help in understanding how proteins, DNA, and RNA function inside cells. A computer has hardware (wires, screens, keyboards, circuits, logic chips, memory, etc.) and also software (computer programs) that

controls the hardware. The hardware is the only thing that actually *does* something. The keyboard and the mouse transmit information from you to the computer; the screen transmits information from the computer back to you; the memory and logic circuits allow the computer to process information; the internet board allows you to send and receive signals from the Web; the speakers play music, and so on. All of this is hardware. On the other hand, stored in the computer's memory are programs that control everything the hardware does. The programs are just patterns of 1's and 0's—they are not made of atoms like the hardware. But these patterns nonetheless determine whether the computer displays your latest homework assignment on the screen, connects you to the Web, or plays music over the speakers. If you had the right hardware, you could design a program that would calculate the surface area of the Moon or one that would launch, fly, and land a rover or spaceship on Mars.

Inside cells, proteins are machines that play the main role of the hardware. They make it possible for the cell to do something, like swim, crawl, divide, or convert sugar into energy. The DNA is like the memory. It stores the instructions and control programs that tell the proteins and other machines what to do. By programming the DNA correctly, you could tell the cell's hardware to go hunting for food, to build an outboard motor to make the cell swim, to attack and eat a foreign cell, to divide and form two new cells, or to continue dividing to build a fish, a frog, or a puppy.

Biological Machinery

The core of any biological organism is the central genetic machinery. The central genetic machinery is the system of DNA, RNA, and proteins that uses the genetic information in the DNA to create new protein machines. It has four main parts: the DNA itself, which contains all the instructions for making a protein; a protein called RNA polymerase, which makes a temporary copy of the DNA called messenger RNA (or mRNA for short); and a very large RNA–protein complex called a ribosome, which uses the mRNA to make a new protein. (See Figure 12.1.)

There are a number of other, smaller parts as well, but these are the main ones. In the next few sections, we will learn more about what proteins are and how the genetic system builds them.

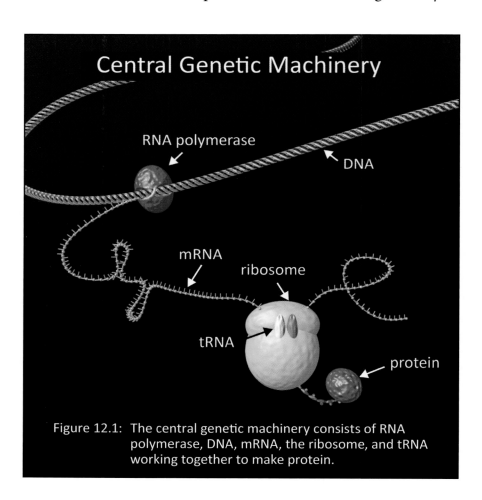

Figure 12.1: The central genetic machinery consists of RNA polymerase, DNA, mRNA, the ribosome, and tRNA working together to make protein.

Chapter 12: Biological Polymers   139

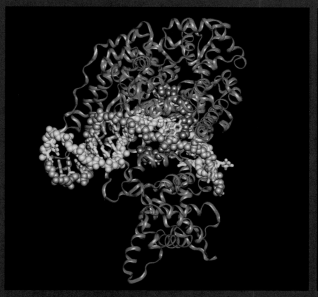

Figure 12.2: RNA polymerase (in red) is a protein machine that makes RNA.

Crystal structure by Yin, Y. W., Steitz, T. A., "The Structural Mechanism of Translocation and Helicase Activity in T7 RNA Polymerase," Cell (Cambridge, MA v116 pp. 393–404, 2004 (Protein Data Bank ID 1S76). Illustration by D. J. Keller

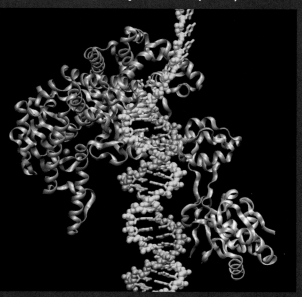

Figure 12.3: DNA polymerase (in blue) is a protein machine that copies DNA.

Crystal structure by Doublie, S., Tabor, S., Long, A. M., Richardson, C. C., Ellenberger, T. "Crystal structure of a bacteriophage T7 DNA replication complex at 2.2 Å resolution," Nature v391 pp. 251-258, 1998 (Protein Data Bank ID 1C57). Illustration by D. J. Keller

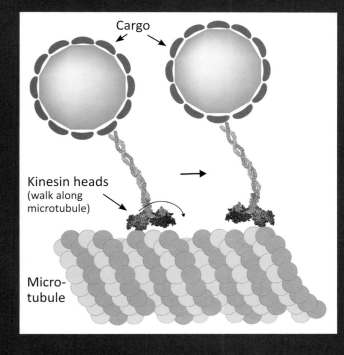

Kinesin

Figure 12.4: Kinesin is a protein motor that moves "cargo" through the cell. It performs this task by "walking" (with the heads) along microtubules.

Kinesin head crystal structure by Kikkawa, M., Sablin, E. P., Okada, Y., Yajima, H., Fletterick, R. J., and Hirokawa, N., "Switch-based Mechanism of Kinesin Motors," Nature v411 pp. 439–445, 2001 (Protein Data Bank ID 1IA0. Illustration by R. W. Keller

12.2 Amino Acid Polymers: Proteins

There are many different kinds of proteins, each one performing a different job inside the cell. Many proteins carry out chemical reactions that help synthesize organic molecules even more efficiently than an organic chemist in the lab! For example, RNA polymerase is a protein that makes the messenger RNA, which is then used by the ribosome to make new protein (see Figure 12.2). Also, every time a cell divides, it needs to make a new copy of its DNA. So a protein called a DNA polymerase makes the copy for the new cell (see Figure 12.3). Some proteins can cut other proteins, DNA, or RNA, and some molecules pump chemicals from inside the cell to outside the cell. Other proteins are part of large and complex protein assemblies, such as those found in bacterial flagella, which spin propellers so cells can swim. Still others, like kinesin, are tiny motors that move molecules from one place to another (see Figure 12.4). Protein motors are everywhere inside every cell doing all of the work that keeps cells living.

Amino Acids

From a chemical point of view, proteins are polymers composed of amino acids. There are 20 "standard" amino acids found in living things. All of the amino acids, except for proline, have a central carbon attached to a carboxylic acid group and an amine group. The central carbon atom is called the α-carbon (pronounced "alpha carbon"). Because they have both an amine and an acid attached to the α-carbon, they are called α-amino acids.

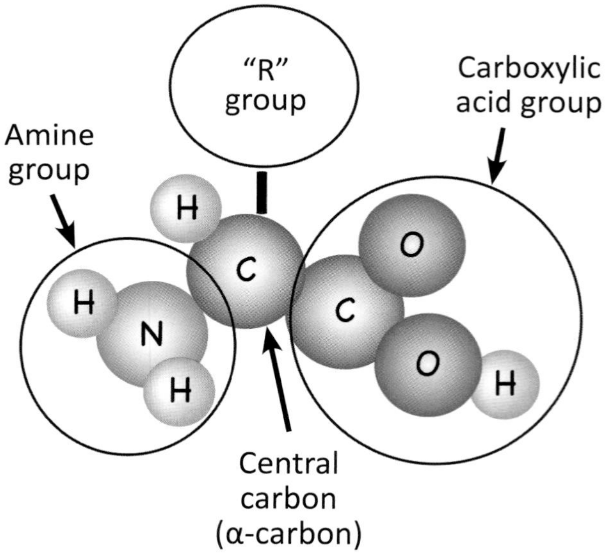

Figure 12.5: An **amino acid** has two functional groups (a carboxylic acid group and an amine group) attached to a central carbon atom called the **α-carbon**.

There is also a third group attached to the α-carbon called the "R" group. The "R" stands for different types of atoms or functional groups. For example, the "R" group in glycine is simply a hydrogen atom, and the "R" group in alanine is a methyl group.

All amino acids have the same "backbone" consisting of the amine group, the α-carbon, and the acid group, but they have different "R" groups. So, when we say that there are 20 different amino acids, we mean that there are 20 different "R" groups. In principle the "R" group could be anything organic, but only these 20 "R" groups are actually used by living things.

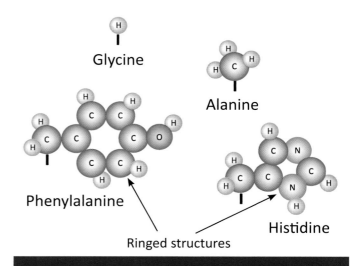

Figure 12.6: The **"R" groups** can be simple (like the "H" of glycine) or more complex (like the benzene ring of phenylalanine).

The "R" groups are what give amino acids different properties. For example, the side chains of some amino acids such as glutamic acid and aspartic acid, have acidic "R" groups. Other amino acids, such as arginine and lysine, have side chain "R" groups that are basic. The side chains of the amino acids alanine, leucine, and valine are *hydrophobic* and do not interact with water. The amino acids asparagine and glutamine contain an amide functional group which is polar, so these side chains are *hydrophilic* and interact with water. The "R" group can also be a ringed structure. For example, the "R" group in phenylalanine has a six-membered benzene ring. Phenylalanine is thus an aromatic molecule in addition to being an amino acid. The "R" group for histidine has a five-membered ring called imidazole. As we will see in the next few sections, the different properties of the "R" groups control the shapes of proteins and make it possible for proteins to do their jobs. (See Appendix for a table of amino acids.)

Peptide Bonds

Proteins are polymers made of amino acids. Most proteins are at least a few hundred amino acids long, and some of the largest proteins are thousands of amino acids long. Proteins are more complicated than the synthetic polymers we learned about in Chapter 11, because instead of one or two monomer types, proteins combine up to 20 different types of amino acid monomers in a single polymer chain.

Amino acids are linked to each other with a peptide bond. A peptide bond is just another name for an amide linkage, exactly like the amide linkages in synthetic polymers such as nylon or KEVLAR®. Amino acids get hooked together by a condensation reaction between the acid group at one end of an amino acid and the amine group at the other end of an adjacent amino acid, as shown in Figure 12.7.

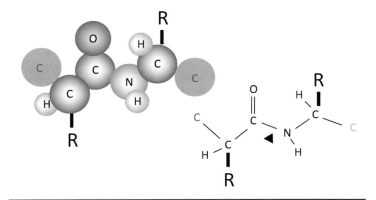

Peptide bond

Figure 12.7: A peptide bond is an amide linkage just like those in the synthetic polymers nylon and KEVLAR®.

Notice that after the reaction the new two-amino-acid molecule still has an amine group at one end and an acid group at the other. This means that more amino acids can be added to either end.

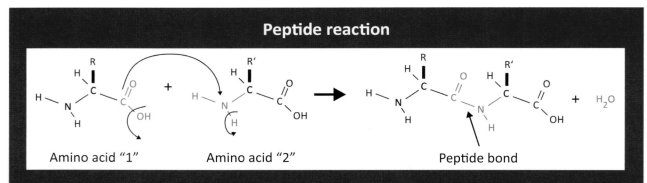

Figure 12.8: In this example, the peptide bond is formed by a condensation reaction between the acid group on "amino acid 1" and the amine group on "amino acid 2."

The ribosome (see Figure 12.22) is the molecular machine that hooks amino acids together. It facilitates, or catalyzes, the formation of the peptide bonds between amino acids. Though it is chemically possible to add amino acids to either end of a growing protein, *ribosomes always add new amino acids to the end with the acid group,* never to the end with the amino group. Any number of amino acids can be linked in any order: Alanine–alanine–alanine–alanine is a possible protein, but so is arginine–glycine–glutamic acid–phenylalanine–alanine.

Protein Primary Structure

The most important thing to remember about proteins is:

> The order of the amino acids in a protein chain
> determines what kind of machine the protein becomes.

Because each protein has a different order of amino acids, one protein can become part of a rotary motor, another a pump for sodium ions, and another an RNA polymerase. Having one or more amino acid monomers out of sequence can destroy the protein's function. The order, or sequence, of amino acids in a polymer is called the primary structure. A short hypothetical protein sequence could be arginine–glycine–glutamic acid–phenylalanine–alanine. Connecting these amino acids with peptide bonds would give us the hypothetical protein shown in Figure 12.9.

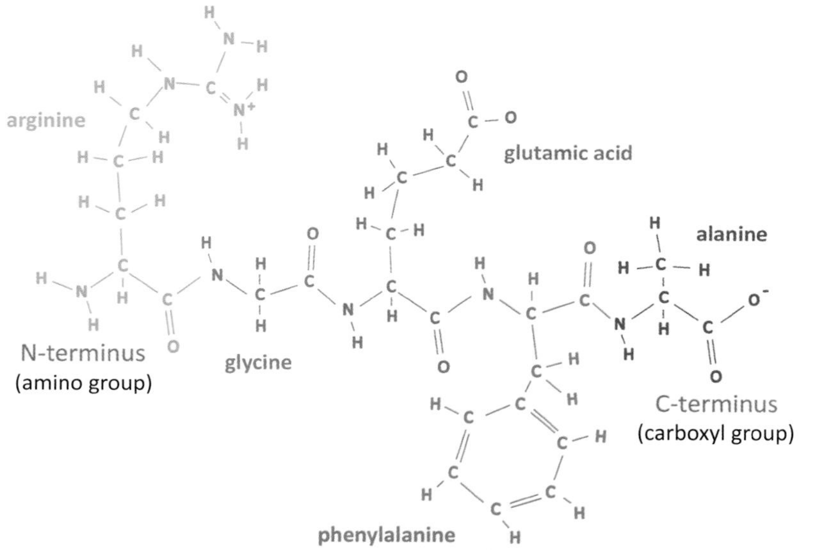

One end of the protein has an amino group and the other end has an acid group, also called a carboxyl group. By convention, the end of the protein with the amino group is called the N-terminus, and the other end, with the carboxyl group, is called the C-terminus. Since protein sequences can be quite long, instead of writing out the entire name of the amino acids, scientists use abbreviations. Biochemists use two types

Polypeptide chain

Figure 12.9: The polypeptide chain arginine-glycine-glutamic acid-phenylalanine-alanine forms a hypothetical protein and can be written as Arg Gly Glu Phe Ala or RGEFA.

of abbreviations: a three-letter abbreviation or a single-letter abbreviation. For example, arginine is given the three-letter abbreviation arg and glycine is given the three-letter abbreviation gly. Using three letter abbreviations, the polypeptide-chain sequence becomes arg–gly–glu–phe–ala, which is much easier to write. In the single-letter abbreviations, arginine is R, glycine is G, glutamic acid is E, phenylalanine is F, and alanine is A. The sequence now becomes RGEFA. The single-letter abbreviations are shorter to write, but require that you remember the single-letter abbreviations, which can be harder to remember. Biochemists use the three-letter abbreviations when the sequence is short and the single-letter abbreviations when it is very long. (See Appendix for amino acid chart and abbreviations.)

Protein Secondary Structure

Once the amino acids are connected together into a long chain, the protein folds up into a compact shape. The sequence of amino acids, or primary structure, determines what shape the protein folds into. The shape of the protein determines what kind of machine the protein will become.

When a protein begins to fold, it first organizes into its secondary structure. There are two basic kinds of secondary structures, called helices and pleated sheets.

Helices are formed when a chain of amino acids, called a polypeptide backbone, twists into a cylindrical coil, much like wrapping a ribbon around a tube. (See Figure 12.10) Notice that a ribbon can be coiled in two directions—left or right. The same is true of protein helices. They can be either right-handed or left-handed. The helices in proteins are right-handed and are called α-helices (pronounced "alpha helices").

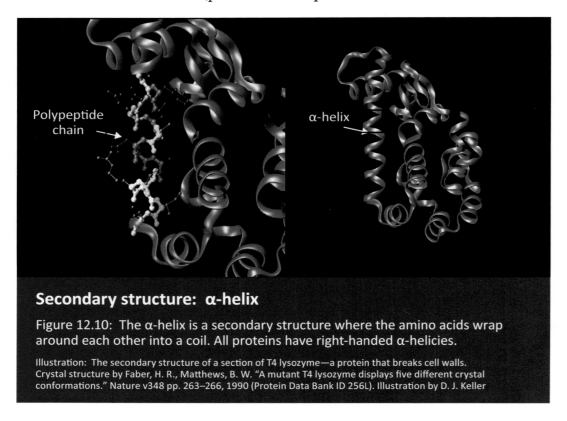

Secondary structure: α-helix

Figure 12.10: The α-helix is a secondary structure where the amino acids wrap around each other into a coil. All proteins have right-handed α-helicies.

Illustration: The secondary structure of a section of T4 lysozyme—a protein that breaks cell walls. Crystal structure by Faber, H. R., Matthews, B. W. "A mutant T4 lysozyme displays five different crystal conformations." Nature v348 pp. 263–266, 1990 (Protein Data Bank ID 256L). Illustration by D. J. Keller

Another pattern that proteins form is called a β-pleated sheet (beta pleated sheet). In a pleated sheet, the polypeptide backbones line up next to each other in a sheet-like structure. (See Figure 12.11)

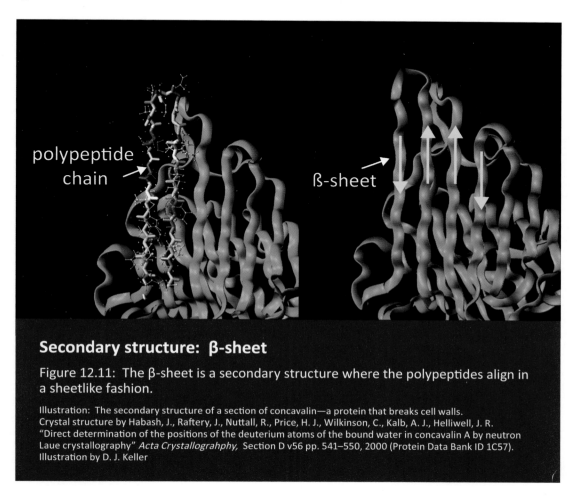

Secondary structure: β-sheet

Figure 12.11: The β-sheet is a secondary structure where the polypeptides align in a sheetlike fashion.

Illustration: The secondary structure of a section of concavalin—a protein that breaks cell walls. Crystal structure by Habash, J., Raftery, J., Nuttall, R., Price, H. J., Wilkinson, C., Kalb, A. J., Helliwell, J. R. "Direct determination of the positions of the deuterium atoms of the bound water in concavalin A by neutron Laue crystallography" *Acta Crystallograhphy,* Section D v56 pp. 541–550, 2000 (Protein Data Bank ID 1C57). Illustration by D. J. Keller

Protein Tertiary Structure

Once the polypeptide backbone folds into one or more secondary structures, the protein further folds into what is called the tertiary structure. The tertiary structure determines the overall shape of the protein and is critical for protein function. A protein with the correct primary and secondary structure but an incorrectly folded tertiary structure will not function properly in the cell.

The secondary structures of a protein fold together to form larger three-dimensional tertiary structures, such as domains and barrels. These structures can function alone or together with other proteins.

Protein domains are clusters of 100–200 amino acid monomers folded into a compact unit. These clusters can contain all of the secondary structures, such as helices and pleated sheets. Protein domains often have specialized functions and hence behave as independent parts of the overall protein machine. For example, there are several protein domains for DNA polymerase (see Figure 12.12) called the palm, thumb, and fingers. The thumb domain and the fingers domain hold a piece of DNA in place, and the palm houses the chemical activity for making new DNA. The fingers open and close adding new nucleotides to the newly formed DNA.

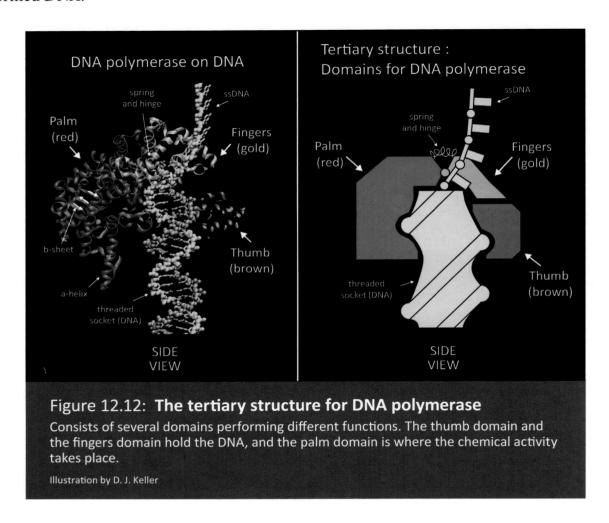

Figure 12.12: **The tertiary structure for DNA polymerase**
Consists of several domains performing different functions. The thumb domain and the fingers domain hold the DNA, and the palm domain is where the chemical activity takes place.

Illustration by D. J. Keller

Barrels are formed primarily by pleated sheets. The pleated sheets line up side by side forming a barrel-shaped structure. The pleated sheets are typically oriented in opposite directions and connected to each other by turns. (See Figure 12.13.)

Chapter 12: Biological Polymers   147

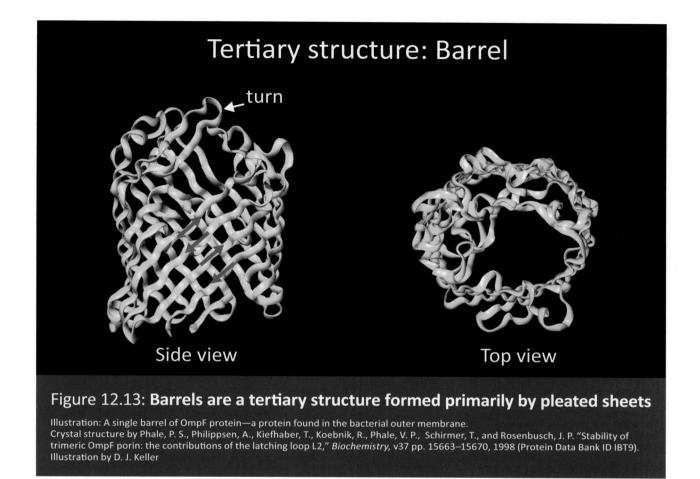

Figure 12.13: **Barrels are a tertiary structure formed primarily by pleated sheets**
Illustration: A single barrel of OmpF protein—a protein found in the bacterial outer membrane.
Crystal structure by Phale, P. S., Philippsen, A., Kiefhaber, T., Koebnik, R., Phale, V. P., Schirmer, T., and Rosenbusch, J. P. "Stability of trimeric OmpF porin: the contributions of the latching loop L2," *Biochemistry*, v37 pp. 15663–15670, 1998 (Protein Data Bank ID IBT9).
Illustration by D. J. Keller

Protein Quaternary Structure

Once a protein has folded into its tertiary structure, it is complete. However, it often happens that two proteins work together to make a single machine. The combination of two or more protein chains functioning together is called quaternary structure. The quaternary structure is yet another level of protein structure complexity. For example, a proteosome (see Figure 12.14) is a large and complex molecular machine that breaks down proteins. It is composed of many protein chains which are arranged into four rings (called α and β) that are stacked, one on top of another. The two β rings are in the center of the proteosome and contain areas (called proteolytic sites) that break apart or degrade proteins.

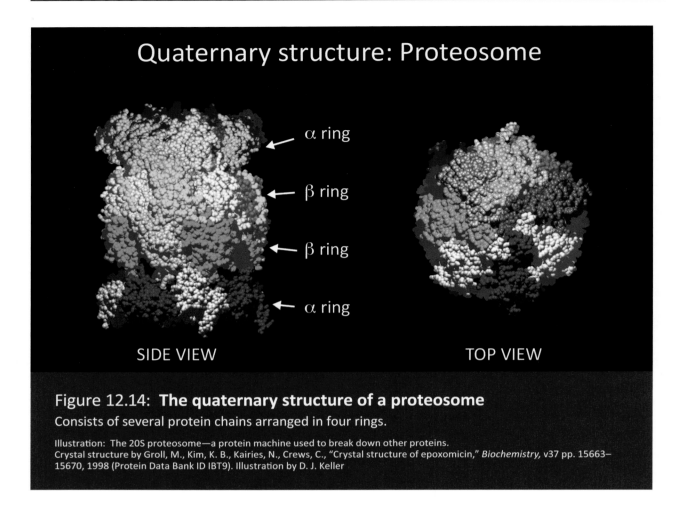

Figure 12.14: **The quaternary structure of a proteosome**
Consists of several protein chains arranged in four rings.

Illustration: The 20S proteosome—a protein machine used to break down other proteins.
Crystal structure by Groll, M., Kim, K. B., Kairies, N., Crews, C., "Crystal structure of epoxomicin," *Biochemistry*, v37 pp. 15663–15670, 1998 (Protein Data Bank ID IBT9). Illustration by D. J. Keller

12.3 Nucleic Acid Polymers

Proteins do most of the actual work inside cells, but DNA and RNA carry the genetic information. DNA and RNA contain the instructions that tell the cell how to grow, when to divide, which proteins to make, how many proteins to make, and when to die. DNA is like the disk drive of a computer—it stores the permanent copy of all the software for the cell. RNA is like the computer's memory chip—it stores temporary copies of instructions from the DNA. Together, DNA and RNA tell the ribosomes which proteins to make and how to make them.

What is the difference between RNA and DNA? And why are DNA and RNA, rather than proteins, used to store information? The answers to both of these questions lie in the *chemical structures* of DNA and RNA. As we will see, DNA and RNA are very similar to each other, and both have structures that are especially well suited to the task of storing information.

Chapter 12: Biological Polymers 149

DNA

Rosalind Franklin (1920-1958 CE) was an English chemist who studied DNA using X-ray crystallography, a technique that uses the scattering of radiation to make an image of large molecules. Francis Crick (1916-2004 CE) and James Watson (1928 born) obtained Franklin's experimental data without her knowledge, and in 1953 they used this data to determine the structure of DNA. The discovery of the structure of DNA marks the birth of modern molecular biology. Before this time very little was known about how cells live or what makes them die. But today we know quite a bit about cellular function and activity.

The three letters "D-N-A" stand for deoxyribonucleic acid. Like a protein, DNA is a polymer made of several different kinds of monomers, and the sequence of these monomers is what allows DNA to do its job. But there are only four different kinds of monomers (instead of the 20 amino acids in proteins), and they are more complicated than amino acids.

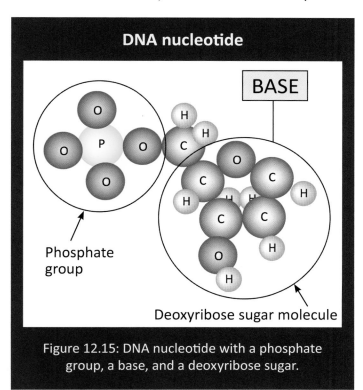

Figure 12.15: DNA nucleotide with a phosphate group, a base, and a deoxyribose sugar.

The monomer unit for DNA is called a nucleotide. A nucleotide has three main parts: a nucleic acid base; a five-membered sugar ring, called a ribose (the "ribo" part of deoxyribonucleic acid); and a phosphate group (which is acidic, like phosphoric acid and is the acid part of deoxyribonucleic acid). The sugar group on DNA is deoxygenated, which means that it is missing an -OH group on one of its carbon atoms. This is the reason that DNA is called *deoxy*ribonucleic acid.

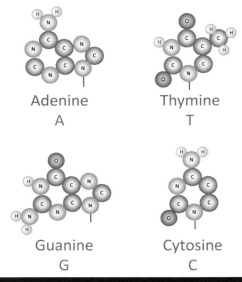

Figure 12.16
The four DNA bases
adenine, thymine, guanine, and cytosine

Genetic Words and the Sequence of DNA

Nucleotides all have the same sugar (deoxyribose) and the same phosphate group, but there are four different kinds of nucleic acid bases: adenine (A), thymine (T), guanine (G), and cytosine (C).

The sequence of bases along the DNA can therefore be written as a sequence of letters. For example, the letters AATCGCAT stand for adenine-adenine—thymine—cytosine—guanine—cytosine—adenine—thymine.

Each of these bases is one letter of the genetic code. The sequence of bases along the DNA chain is exactly like words on a page or like the 1's and 0's in a computer memory. They spell out the genetic information that a cell needs to make proteins and to control its network of machines.

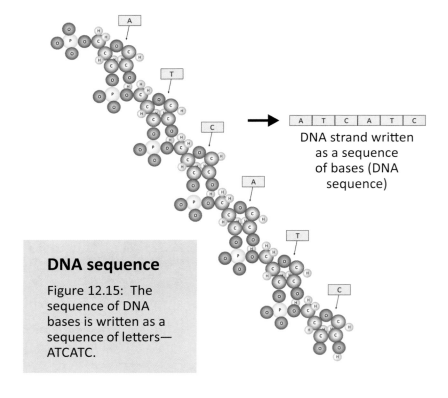

DNA sequence

Figure 12.15: The sequence of DNA bases is written as a sequence of letters—ATCATC.

DNA strand written as a sequence of bases (DNA sequence)

The Double Helix

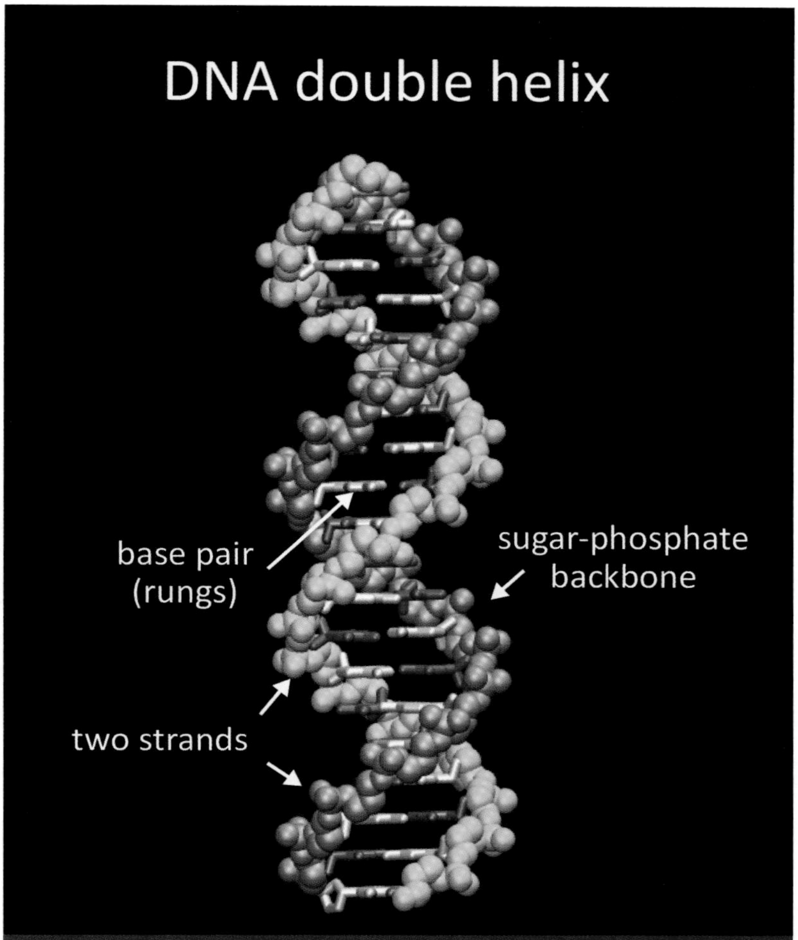

The full DNA polymer usually contains two DNA chains and is called double-stranded DNA, or dsDNA for short. Since it has two DNA chains, it has two copies of the genetic information. The two chains are twisted around each other with the nucleic acid bases in the middle. The bases on one strand stick edge to edge to the bases on the other strand by means of their hydrogen bonds. So the whole structure is like a ladder that has been twisted, with the bases like the rungs of the ladder and the sugar-phosphate backbone like the rails.

The stuck-together bases in the middle of the double helix are called base pairs. An A base on one strand is always paired with a T base on the other strand. Likewise, a G base is always paired with a C base. So the "rungs" in the ladder are always A–T or G–C. In this way it is always possible to tell, from looking at one strand, what the sequence is on the other strand.

Figure 12.17: A DNA double helix
Two chains of nucleic acid polymers are wrapped around each other.
Illustration by D. J. Keller

The double helix structure of DNA is important for its function, which is the safe storage and accurate transmission of genetic information. That is, DNA has to be able to safely store and express the correct codes that the cell needs to make proteins, metabolize nutrients, grow, and divide. The code is found in the bases which are tucked safely inside the double helical coil. Here, they are not easily removed or degraded. In this way, DNA is able to protect the important information that the cell needs to live.

12.4 RNA

The letters "R-N-A" stand for ribonucleic acid. RNA is made from nucleotides with a ribose sugar, a phosphate group, and a nucleic acid base, just like DNA, but with two important differences. First, RNA contains two hydroxyl groups on its ribose sugar and is not "deoxygenated," like DNA. This is why RNA is called "ribonucleic acid" instead of "deoxyribonucleic acid."

Second, RNA does not have a thymine (T) base, but instead uses a base called uracil (U) in place of thymine. These two small differences make RNA a significantly different molecule than DNA. The base pairing in RNA is the same as in DNA, except A forms a base pair with U instead of with T.

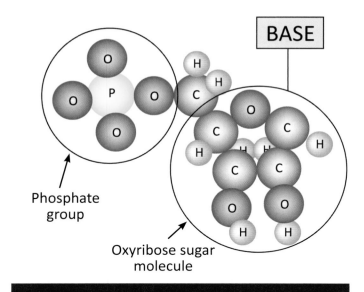

RNA nucleotide

Figure 12.18: An RNA nucleotide is composed of an oxygenated ribose sugar, a phosphate group, and a base.

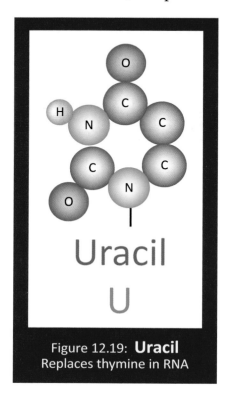

Figure 12.19: **Uracil** Replaces thymine in RNA

RNA is usually found in complex structures, rather than in a double helix like DNA. RNA can fold back on itself, base pairing in certain regions. Because of this, RNA has many different functions inside cells. The ribosome—the machine that makes proteins—is mostly made of RNA. Another important class of molecules is transfer RNA (tRNA)—smaller RNA molecules that play an important role in the synthesis of new proteins.

But perhaps the most important form of RNA is messenger RNA (mRNA)—the temporary copy of a gene that tells the ribosome which amino acids to use when constructing a protein. Now that we know the basics about DNA, RNA, and proteins, we are ready to learn how they work together to use genetic information to make new proteins.

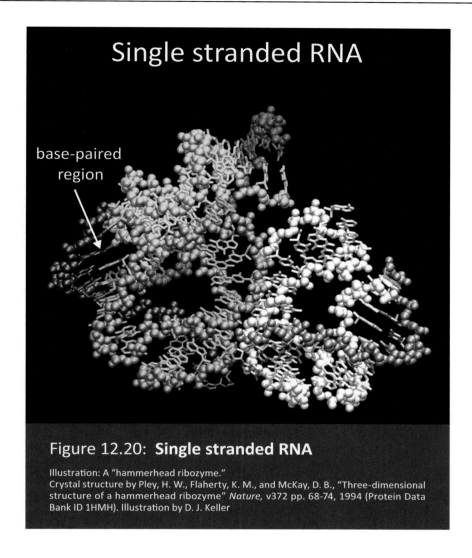

Figure 12.20: **Single stranded RNA**

Illustration: A "hammerhead ribozyme."
Crystal structure by Pley, H. W., Flaherty, K. M., and McKay, D. B., "Three-dimensional structure of a hammerhead ribozyme" *Nature,* v372 pp. 68-74, 1994 (Protein Data Bank ID 1HMH). Illustration by D. J. Keller

12.5 Building Biological Machines

Cells need new proteins all the time. Whenever a cell has a big job to do—dividing, crawling, hunting for food—it must make a host of new protein machines to carry out the job. Earlier in the chapter, we briefly described the central genetic machinery, which consists of the DNA, an RNA polymerase, a messenger RNA, and the ribosome (plus numerous other parts). To make new proteins, cells send orders, (like "Make protein X!") to the central genetic machinery. (You may wonder, "How does a cell send an order to its own DNA?" That is an excellent question, but too complicated for us here! For now we will just accept that it can.)

The genetic letters in the DNA carry the code for the amino acids in the protein that is to be made, so the DNA contains the basic plan for the protein. The genetic code itself is found in Table 1. Using this table, you can "read" the genetic information in the DNA. At first glance, the sequence of letters in a strand of DNA looks random and meaningless, however it is really a series of three-letter codons. Most of the codons are for amino acids, but some are also "start" and "stop" codons. As we saw earlier, a hypothetical DNA molecule with the sequence AGA–TGA–ACC–CTT codes for the amino acids serine–threonine–tryptophan–glutamic acid. How does this DNA sequence get copied and used to make a new protein?

The first step in making a new protein is called transcription. Transcription begins when an RNA polymerase binds to the DNA. The job of the RNA polymerase is to copy (or transcribe) the DNA to make a piece of RNA with a complementary sequence to the DNA. In this way, the instructions for the protein have been transferred to the RNA. A complementary sequence means that the RNA copy of the DNA is made with complementary base pairs for RNA.

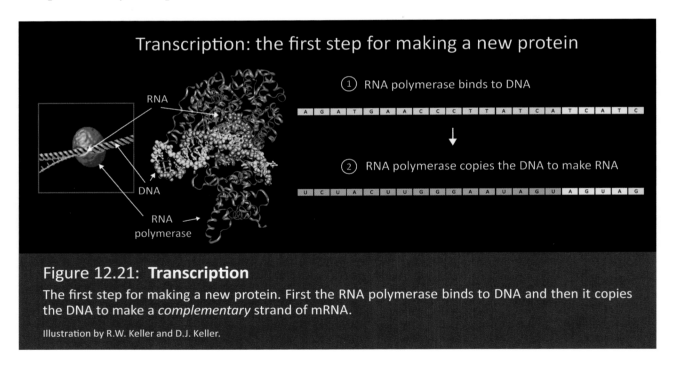

Figure 12.21: Transcription
The first step for making a new protein. First the RNA polymerase binds to DNA and then it copies the DNA to make a *complementary* strand of mRNA.

Illustration by R.W. Keller and D.J. Keller.

So, the A nucleic acid residues in the DNA are U nucleic acid residues in the RNA. The T nucleic acid residues in the DNA are A nucleic acid residues in the RNA. The G nucleic acid residues are C nucleic acid residues in the RNA, and the C nucleic acid residues in the DNA are G nucleic acid residues in the RNA.

Table 1	Amino acid codons (RNA)			
Alanine (Ala)	**Arginine (Arg)**	**Asparagine (Asn)**	**Aspartic acid (Asp)**	**Cysteine (Cys)**
GCU GCC GCA GCG	AGA AGG CGU CGC CGA CGG	AAU AAC	GAU GAC	UGU UGC
Glutamic acid (Glu)	**Glutamine (Gln)**	**Glycine (Gly)**	**Histidine (His)**	**Isoleucine (Ile)**
GAA GAG	CAA CAG	GGU GGC GGA GGG	CAU CAC	AUU AUC AUA
Leucine (Leu)	**Lysine (Lys)**	**Methionine (Met)**	**Phenylalanine (Phe)**	**Proline (Pro)**
CUU CUC CUA CUG UUA UUG	AAA AAG	AUG	UUU UUC	CCU CCC CCA CCG
Serine (Ser)	**Threonine (Thr)**	**Tryptophan (Trp)**	**Tyrosine (Tyr)**	**Valine (Val)**
UCU UCC UCA UCG AGU AGC	ACU ACC ACA ACG	UGG	UAU UAC	GUU GUC GUA GUG

STOP
UAA UAG UGA

The RNA polymerase will only bind at special control sites on the DNA called promoters (a sequence like GAGGCTATATATTCCCCAGGGCTCAGCCAGTGTCTGTAACA), so every gene must have a promoter in front of it. After the RNA polymerase binds, it begins to crawl along the DNA. As it crawls, it unwinds the DNA double helix, exposing the genetic letters inside. Then it copies each DNA letter by adding a new RNA monomer to a growing chain of RNA. This is the messenger RNA (mRNA) mentioned earlier.

Once the DNA has been accurately copied into a messenger RNA molecule, it can be used to make a protein. This process is called translation. The messenger RNA carrying the DNA code binds to a ribosome. Unlike the DNA, the RNA copy is not a double helix, so the bases of the RNA are exposed and can easily be read by the ribosome. The ribosome reads (or translates) the mRNA three letters at a time with the help of transfer RNA molecules, or tRNAs for short. Transfer RNAs are like little plugs, each with an amino acid on its end. The "plug" end of each tRNA has a set of three bases, called an anticodon, that must match the codon on the mRNA.

The mRNA molecule binds to the inside of the ribosome. Once it is securely fastened, tRNA molecules carrying the amino acids for each new amino acid align with the three-letter codons on the mRNA. For each tRNA, the ribosome links the amino acid it is carrying into the growing chain of amino acids that make up the new protein. As the protein chain grows, it folds into its proper shape (secondary structure, tertiary structure, and with other proteins, quaternary structure) and becomes a new protein machine. (See Figure 12.22)

12.6 Summary

- Two important polymers found in living systems are amino acid polymers (proteins) and nucleic acid polymers (DNA and RNA).
- Amino acids are linked by peptide bonds between the carboxyl group on one amino acid and the amine group on the adjacent amino acid.
- The order, or sequence, of amino acids determines the primary structure of proteins.
- The secondary structure of a protein is the way that the polypeptide folds into coils and sheets.
- The tertiary structure of a protein is the way that the coils and sheets of the secondary structure fold into a complex three-dimensional shape.
- The structure of a protein is necessary for its function.

Chapter 12: Biological Polymers 157

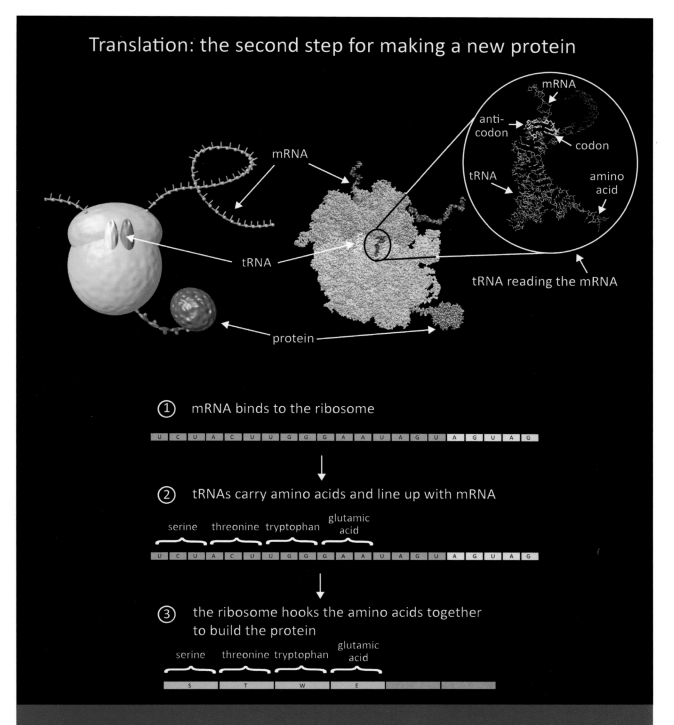

Figure 12.22: **Translation**
The second step for making a new protein.

Ribosome structure from Jenner, L., Romby, P., Rees, B., Schulze-Briese, C., Springer, M., Ehresmann, C., Eshresmann, B., Moras, D., Yusupova, G., Yusupov, M. "Translational operator of mRNA on the ribosome: hw repressor proteins exclude ribosome binding" *Science*, v308 pp. 120-123, 2005 (Protein Data Bank ID 1YL4). Illustration by R. W. Keller and D. J. Keller

- DNA (deoxyribonucleic acid) and RNA (ribonucleic acid) are the two biological polymers that carry, store, and transmit information inside cells for growth, reproduction, and metabolism.

- The genetic information in DNA or RNA is given by the base pair sequence of the four bases, which is A,T,C and G in DNA, and A, U, C, and G in RNA.

- DNA is a double helix formed from two chains of DNA monomers twisted around each other like a twisted ladder with the bases base-paired across the middle. The structure of DNA allows the genetic information to be safely stored.

- RNA polymerase is a protein that "reads" the DNA in order to make a temporary copy of its genetic information out of RNA. The temporary copy is called messenger RNA, or mRNA.

- Proteins are made by a protein machine called a ribosome. The ribosome reads the three-letter combinations of the genetic code from the messenger RNA and creates a protein with the matching sequence of amino acids.

12.7 Some Things to Think About

- Explain the functions of proteins, DNA, and RNA.
- Name and describe the four structures of proteins.
- Why is the order of amino acids in a protein chain important?
- Why are DNA and RNA important to the body?
- Describe the double helix structure of DNA.
- Why is molecular recognition important to DNA and RNA?
- What are the differences between DNA and RNA?
- What is messenger RNA?
- Describe the processes of transcription and translation.

Appendix

Amino Acid "R" Groups

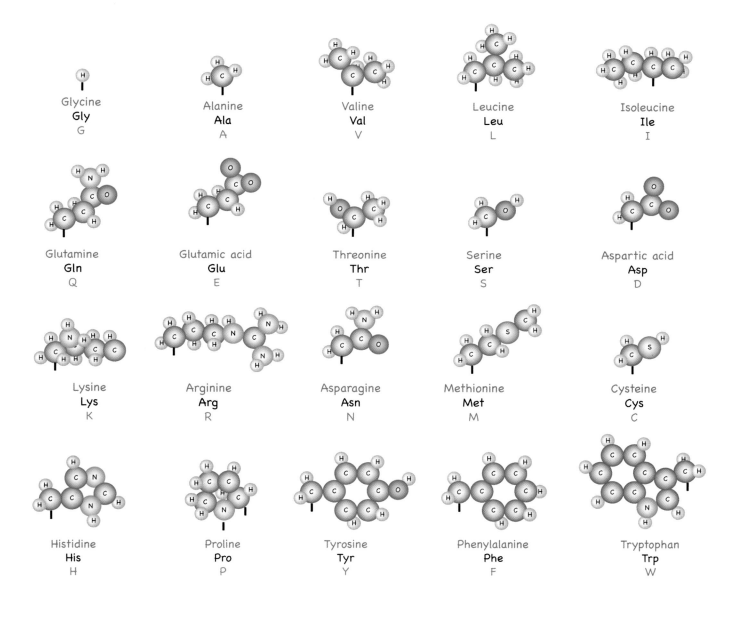

Glossary-Index

[Pronunciation Key at end]

α ('al-fə) • the Greek letter alpha used in scientific terminology, 140

α-amino acid ('al-fə ə-'mē-nō 'a-səd) • an amino acid having both an amine and an acid attached to the α-carbon, 140

α-carbon ('al-fə 'kär-bən) • the central carbon atom of an amino acid that bonds to both the carboxylic acid group and the amine group, 140-141

α-helix ('al-fə 'hē-liks), [pl., **α-helices** ('al-fə 'he-lə-sēz)] • a helix in a protein that is twisted in a right-handed direction, 144

A • one-letter abbreviation for alanine and for adenine, 143

acid ('a-səd) • any of a number of molecules whose pH in water is below 7; will react with a base to form a salt, 52-59, 62-70, 94, 110, 114-116, 125, 131-132, 133, 140, 141, 142, 143, 149

acid-base reaction • see reaction, acid-base

acid, citric • see citric acid

acid indigestion ('a-səd in-dī-'jes-chən) • a physical condition that results when the stomach has produced too much acid and the excess acid causes pain, 63

addition reaction • see reaction, addition

adenine ('a-də-nēn) [abbrev., **A**] • one of the four nucleic acid bases that make up DNA and RNA, 150

aerosol ('er-ə-säl) • a heterogeneous mixture made of liquids or solids mixed into a gas, 91

alanine ('a-lə-nēn) [abbrev., **ala** or **A**] • an amino acid that has a methyl "R" group and is hydrophobic, 140, 141, 142, 143, 155

alchemist ('al-kə-mist) • one of the early experimenters who tried to turn one kind of matter into another kind of matter, such as lead into gold, 7, 8

alchemy ('al-kə-mē) • the philosophy that one kind of matter can be changed into a different type of matter, 7

alcohols ('al-kə-hôlz) • a class of organic molecules that contain an -OH functional group attached to a carbon atom, 94, 99-100, 102-103, 110, 114-115, 127, 130, 131, 133

aldehydes ('al-də-hīdz) • a class of organic compounds that contain a -CH=O functional group, 110, 115-116

alembic (ə-'lem-bik) • a type of lab glassware used in distillation, 15

alkali ('al-kə-lī) **metal** • one of a group of elements that react with many different atoms and molecules, 28

alkanes ('al-kānz) • a class of organic compounds that contain only carbon and hydrogen with single bonds, 110, 111-112

alkenes ('al-kēnz) • a class of organic compounds that contain one or more double bonds between adjacent carbon atoms, 110, 111, 112, 125

alkynes ('al-kīnz) • a class of organic compounds that contain one or more triple bonds between adjacent carbon atoms, 110, 111, 112-113

amide ('a-mīdz) • a molecule in the class of organic compounds that contain a -C=O-NH- functional group, a molecule made by linking an acid and an amine, 132, 133, 141, 142

amine ('ə-mēn) **group** • a functional group in an amino acid, 140-142

amines ('a-mēnz) • a class of organic compounds that contain a -NH₂ functional group attached to a carbon atom, 110, 114-115, 132, 133, 140-142

amino acid (ə-'mē-nō 'a-səd) • a molecule that has two functional groups—a carboxylic acid group and an amine group—attached to a central carbon atom called the α-carbon), 140-144, 146, 149, 152, 154, 155, 156

amino acid polymer (ə-'mē-nō 'a-səd 'pä-lə-mər,) • a protein, 136, 140-148

amu (pronounced "A" "M" "U") • abbreviation for atomic mass unit, 67

amylopectin (a-mə-lō-'pek-tən) • a starch made by plants, 81, 119, 120

amylose ('a-mə-lōs) • a starch made by plants, 81, 119

analytical balance • see balance, analytical

analytical chemist • see chemist, analytical

Anaximander (ə-'nak-sə-man-der) • [611-547 BCE] a Greek philosopher who asked about what things are made of; came up with the undefined idea of the "boundless," 3, 6

Anaximenes (a-nak-'si-mə-nēz) • [c. 585-525 BCE] a Greek philosopher who believed that air is the basic substance of matter, 4, 6

antacid (ant-'a-səd) • a basic substance that is taken to reduce the pain of indigestion by neutralizing stomach acid, 63

anticodon (an-tē-'kō-dän) • a set of three bases on transfer RNA that connects with a codon of three amino acids on messenger RNA, 156

aqua regia ('ä-kwə 'rē-jə) • [L., royal water] a mixture of nitric acid and hydrochloric acid that can dissolve gold, 52

aqueous ('ā-kwē-əs) • having to do with water; in chemistry, any solution whose solvent is water, 54, 58, 97, 101

arg • three-letter abbreviation for arginine, 143, 155

arginine ('är-jə-nēn) [abbrev., **arg** or **R**] • an amino acid that is basic, 141, 142, 143, 155

aromatics (a-rə-'ma-tiks) • a class of organic molecules that contain a benzene ring, 110, 111, 113, 141

Arrhenius, Svante (ə-'rē-nē-əs, svän-tā) • [1859-1927 CE] Swedish chemist; showed that acids produce hydrogen ions (H+) in water and bases produce hydroxide ions (OH-) in water, 53

Arrhenius acid (ə-'rē-nē-əs 'a-səd) • any molecule that releases a hydrogen ion (H+), 54, 55

Arrhenius (ə-'rē-nē-əs) **base** • any molecule that releases a hydroxide ion (OH-), 54, 55

asparagine (ə-'sper-ə-ˌjēn) • an amino acid that contains an amide functional group and is hydrophilic, 141, 155

aspartic acid (ə-'spär-tik 'a-səd) • an acidic amino acid, 141, 155

assay ('a-sā) • a process done to analyze a substance for the presence, absence, or quantity of one or more components, 63

atom ('a-təm) • [Gr., *atomos*, uncuttable] the smallest unit of matter; an element, 5-6, 8, 9-10, 13, 21, 24-29, 32-40, 43-49, 52, 53-54, 55, 62-63, 67-68, 82, 85--93, 96, 97, 108-119, 140, 149

atomic (ə-'tä-mik) **mass** • the mass of an atom; comes mainly from the protons and neutrons, 25, 67

atomic (ə-'tä-mik) **mass unit (amu)** • a unit of measurement for the mass of an atom; 1 amu is equal to 1/12th the mass of a carbon atom, 67, 68

atomic (ə-'tä-mik) **number** • the number of protons in the nucleus of an atom, 26-28, 85

atomic (ə-'tä-mik) **weight** • the weight of an atom, which is very close to the combined weight of an atom's protons and neutrons, 10, 27-28, 67

Avogadro, Amedeo (a-və-'gä-drō, äm-ə-'dā-ō) • [1776-1856 CE] Italian chemist and physicist; developed Avogadro's constant to calculate the number of molecules in a given volume, 68

Avogadro's (a-və-'gä-drōz) **constant** • a quantity used to calculate the number of molecules in a given volume; also called Avogadro's number, 68

β-pleated sheet ('bā-tə 'plē-təd 'shēt) • in a protein, a secondary structure where the polypeptide backbones line up next to each other in a sheet-like structure, 145

balance ('ba-lən(t)s) • an instrument that measures the mass of an object or substance, 13, 18-20

balance, analytical ('ba-lən(t)s, a-nə-'li-ti-kəl) • a balance housed in glass; used for making accurate measurement, 20

balance, double-pan • a balance with two platforms, one on either side of a pivot point; also called a Roberval balance; 19

balance, Roberval ('ba-lən(t)s, 'rä-bər-vəl) • a balance with two platforms, one on either side of a pivot point; a double-pan balance, 18

barrel • in a protein, a barrel shaped tertiary structure formed primarily by pleated sheets lined up side by side, 145, 146-147

base • a substance with a pH greater than 7; will react with an acid to form a salt, 52-59, 62-70, 75, 94, 149-156

base pairs • two nucleic acid bases that are the "rungs" in the middle of a double helix that connect two strands of DNA by means of hydrogen bonds, 151, 152, 154

base pairs, complementary (käm-plə-'men-tə-rē) • base pairs in RNA, 154

beaker ('bē-kər) • a container with a wide mouth, flat bottom, and a spout for pouring; 16

Beckman, Arnold • [1900-2004 CE] American chemist; inventor of the first successful pH meter, 59

benzene ('ben-zēn) • an aromatic hydrocarbon with a ring of six carbon atoms and six hydrogen atoms shaped in a hexagon with alternating single and double bonds, 113, 141

biochemist (bī-ō-'ke-mist) • a scientist who studies chemistry as it applies to biological organisms, 13

Bohr, Niels ('bôr, 'nēlz) • [1885-1962 CE], Danish physicist; created a model of the atom in which high and low energy electrons have different, fixed orbits, 34

bond • the attachment between two atoms in a molecule, 21, 32-40, 45, 80, 82, 86, 87, 93, 108, 111, 112, 113, 115, 120-121, 124-130, 141-142, 143, 151

bond, covalent (kō-'vā-lənt) • a molecular bond that has shared electrons, 38, 124, 126, 127

bond, ionic (ī-'ä-nik) • a molecular bond that has unshared electrons, 39, 93, 94

Boyle (boil), **Sir Robert** (1627-1691 CE) • an Irish chemist and philosopher who did experiments to prove or disprove his ideas; figured out fundamental gas laws, 8

branched polymer • see polymer, branched

bunsen burner • a small gas burner used to heat substances in a chemistry lab, 13

C • abbreviation for cytosine, 150

carbohydrate (kär-bō-'hī-drāt) • a molecule containing both carbon and water; the most abundant class of biological molecules, 75, 79-80, 88, 115, 117-120, 136

carbon ('kär-bən) • an element with 6 protons, 6 electrons, and 6 neutrons that is able to form more kinds of molecules than any other element; the focus of organic chemistry, 26, 27, 35, 37, 38, 40, 67, 68, 79, 85, 86, 87, 108-122, 127, 128, 129, 140-141, 149

carbonation (kär-bə-'nā-shən) • the process of adding carbon dioxide gas to a liquid, 8

carbonyl ('kär-bə-nil) • a carbon atom that is double bonded to an oxygen atom [-C=O], 115, 127

carboxyl (kär-'bäk-sil) **group** • an acid group in an amino acid, 143

carboxylic (kär-bäk-'si-lik) **acid group** • a functional group [-COOH] in an amino acid, 140

catalyze ('ka-tə-līz) • to facilitate a chemical reaction, typically speeding it up, 142

cellulose ('sel-yə-lōs) • a polysaccharide that is the main ingredient of wood, cotton, flax, wood pulp, and other plant fibers, 81, 82, 115, 118

central genetic machinery • the system of DNA, RNA, and proteins that uses the genetic information in the DNA to create new protein machines, 138, 153

chain termination • a chemical reaction that stops a polymer chain from further growth, 129

charge, electric • a fundamental property of matter that begins at the atomic level with electrons having a negative charge, protons a positive charge, and neutrons no charge, 24, 39, 53, 92-93

chemical property ('ke-mi-kəl 'prä-pər-tē) • the property of an atom or molecule that results in chemical reactions, 28-29, 91, 96-97

chemical reaction ('ke-mi-kəl rē-'ak-shən) • occurs whenever bonds between atoms and molecules are created or destroyed, 24, 28, 35, 43-49, 52, 54, 62-70, 72, 75, 78, 91, 96, 97, 101, 125, 127-129, 131-133, 140, 142

chemist ('ke-mist) • a scientist who works in the field of chemistry, 13

chemist, analytical ('ke-mist, a-nə-'li-ti-kəl) • a scientist who studies the composition and structure of matter, 13

chemist, organic ('ke-mist, ôr-'ga-nik) • a scientist who studies carbon-containing compounds, 13

chemist, physical ('ke-mist, 'fi-zi-kəl) • a scientist interested in the physical structure and motion of atoms and molecules, 13

chemistry ('ke-mə-strē) • the field of science that studies the composition, structure, and properties of matter, 2

chemistry, inorganic ('ke-mə-strē, in-ôr-'ga-nik) • the branch of chemistry that studies molecules that do not contain carbon, 108

chemistry, organic ('ke-mə-strē, ôr-'ga-nik) • the branch of chemistry that studies carbon-containing molecules, 108-122

chitin (kī-tən) • the main structural polysaccharide found in the exoskeletons of insects, spiders, and crustaceans, 118, 119

cholesterol (kə-'les-tə-rōl) • the most common steroid in animals, 121-122

chromatograph (krō-'ma-tə-,graf), **gas** • an instrument used to analyze mixtures of molecules that are volatile and can be turned into a gas, 20-21

chromatography (krō-mə-'tä-grə-fē) • the process by which components of a mixture are separated using differences in mobility—the difference in how fast each component moves through a given medium, 100, 103-104

chromatography, column (krō-mə-'tä-grə-fē, 'kä-ləm) • a type of chromatography in which a liquid is passed over a column of silicon beads, 104

chromatography, liquid (krō-mə-'tä-grə-fē, 'li-kwəd) • a method of separating components of a mixture of molecules dissolved in a liquid, 104

chromatography (krō-mə-'tä-grə-fē), **paper** • a type of chromatography that uses paper as the stationary phase, 104

citric acid ('si-trik 'a-səd) • an acid found in certain fruits, such as grapefruit and lemon, 53

codon ('kō-dän) • a three-letter combination for the genetic code; a set of three nucleotides in a strand of DNA that is an amino acid or gives a start or stop command, 154, 155, 156

coil, random • a linear, long-chain polymer in which the chain of monomer units is rolled up into a coil, 126

colloid ('kä-loid) • a heterogeneous solution that may look like a homogeneous solution but has particles larger than individual molecules, 90-91, 96

column chromatography • see chromatography, column

combination reaction • see reaction, combination

complementary base pairs • see base pairs, complementary

Glossary-Index 163

complementary sequence see sequence, complementary

compound ('käm-pound) • a chemically bonded pure substance composed of only one kind of molecule, 85-87, 88, 91, 93, 97

concentrate ('kän-sən-trāt) • in chemistry, to make less dilute, 55

concentrated ('kän-sən-trā-təd) • in chemistry, a solution that contains many units in a given volume, 55, 57, 63, 66

concentration (kän-sən-'trā-shən) • in chemistry, the number of units in a given volume, 55, 56, 58, 63, 64, 66

condensation polymer • see polymer, condensation

condensation reaction • see reaction, condensation

condense (kən-'den(t)s) • to make something more dense or compact; to change a substance from a gaseous state to a liquid or solid state, 4, 102-103

condenser (kən-'den(t)-sər) • an apparatus in which a gas is changed into its liquid state, 15, 102

copolymer (kō-'päl-ə-mər) • a polymer made of more than one type of monomer, 125, 127

core • in an atom, the center made up of protons and neutrons; also called the nucleus, 24

covalent bond • see bond, covalent

Crick, Francis • (1916-2004 CE) a British molecular biologist; with James Watson, determined the structure of DNA, 149

cross-link • a covalent bond between two polymer chains, 127, 129, 136

crucible ('krü-sə-bəl) • a container that can be used with high heat, 14

crucible ('krü-sə-bəl) **tongs** • a scissors-like metal tool that is used to grasp a hot crucible, 14

C-terminus ('sē 'tər-mə-nəs) • an acid group (carboxyl group) on one end of a protein molecule, 143

cytosine ('sī-tə-sēn) [abbrev., **C**] • one of the four nucleic acid bases that make up DNA and RNA, 150

Dalton, John • [1766-1844 CE] proposed that all matter is made of atoms; created the first table of elements, 10

decomposition reaction • see reaction, decomposition

Democritus (di-'mä-krə-təs) • [c. 460-370 BCE] • a Greek philosopher who named the atom and believed it was the smallest unit of matter, 5, 6, 9

dendrimer ('den-drə-mər) • an extensively branched polymer, 126

deoxygenated (dē-'äk-si-jə-nā-təd) • a molecule that has had oxygen removed; in DNA, the sugar group that is missing the -OH group on one of its carbon atoms, 149, 152

deoxyribonucleic acid (dē-'äk-si-rī-bō-nü-klē-ik 'a-səd) • DNA (see DNA), 149

Descartes, René (dā-'kärt, rə-'nā) • [1596-1650 CE] French philosopher; theorized that molecules are held together with hooks and eyes, 36

de Villa Nova, Arnaldus ('də 'vil-ə 'nō-və, 'är-näl-dəs) • [circa 1235-1311 CE] Spanish alchemist; first one to use litmus paper to test for acids and bases, 54

digital scale • see scale, digital

dilute (də-'lüt) • in chemistry, a weak solution that contains few units in a given volume, 55

disaccharide (dī-'sa-kə-rīd) • a molecule made of two single sugar molecules, 80, 117, 118

displacement reaction • see reaction, displacement

distillation (dis-tə-'lā-shən) • a process that separates a liquid mixture into its different individual components by means of evaporation and condensation, 15, 17, 100, 102-103

DNA ('dē 'en 'ā) [**deoxyribonucleic acid** (dē-'äk-si-rī-bō-nü-klē-ik 'a-səd)] • a polymer that provides long-term storage of genetic information in cells; contains all of the instructions for making proteins, 136, 137, 138, 139, 140, 146, 148-151, 152, 153-154, 156

DNA, double-stranded [abbrev., **dsDNA**] • the full DNA polymer that has two DNA chains joined by nucleic acid bases, 151

DNA polymerase ('dē 'en 'ā pə-'lim-ə-rās) • a protein that makes a copy of the DNA to be used in making a new cell, 139, 140, 146

domain (dō-'mān) • in a protein, a large three-dimensional tertiary structure of clusters of 100–200 amino acid monomers folded into a compact unit, 145, 146

domain, protein • a cluster of 100–200 amino acid monomers folded into a compact unit, 146

double helix ('hē-liks) • the ladder-like structure of DNA— two chains of nucleic acid polymers that are twisted around each other and are connected by nucleic acid bases, 151, 152, 156

double-pan balance • see balance, double-pan

double-stranded DNA • see DNA, double-stranded

dsDNA • abbreviation for double-stranded DNA, 151

E • the one-letter abbreviation for glutamic acid, 143

elastomer (i-'las-tə-mer) • a polymer that has the ability to stretch and spring back to its original shape, 128

electrical resistance (i-'lek-tri-kəl ri-'zi-stən(t)s) • how well electricity moves through a material, 58

electrolyte (i-'lek-trə-līt) • an ion that can conduct electricity when dissolved in water, 53, 58

electron (i-'lek-trän) • one of the three fundamental particles that make up an atom; found outside the nucleus; has a negative charge and almost no atomic mass, 21, 24-25, 27-29, 32, 33-36, 37-40, 54, 55, 85, 93, 128

element ('e-lə-mənt) • the smallest unit of matter; an atom, 7, 8, 9-10, 24, 25-29, 76, 79, 85-87, 97, 108

element, neutral ('e-lə-mənt, nü-trəl) • an element that has no charge, 24

elemental (e-lə-'men-təl) • referring to a substance composed of only one type of atom, 86

Empedocles (em-'pe-də-klēz) • [c. 490-430 BCE] a Greek philosopher who came up with a periodic table of earth, air, fire, and water, which he said made up all matter, 4, 6

emulsion (i-'məl-shən) • a type of heterogeneous colloid where a liquid is mixed together with a liquid, 91, 95, 96

emulsion (i-'məl-shən), **solid** • a heterogenous colloid made of a liquid mixed into a solid, 91

enzyme ('en-zīm) • a special protein that breaks the long chains of glucose in starches into individual glucose molecules, 37, 81, 82

epoxy resin • a hard synthetic polymer, 13

Erlenmeyer flask • see flask, Erlenmeyer

erythrose (i-'rith-rōs) • a monosaccharide that is a tetrose with four carbon atoms, 117

essence ('e-sən(t)s) • the true nature of a thing, 3

ester ('es-tər) • one of a class of molecules with the functional group -C=O-O- that provide an easy way to hook two big molecules together, molecules made by hooking together an acid and an alcohol, 131, 133

ethene ('e-thēn) • another name for ethylene, 112, 125, 128, 129

ethylene ('e-thə-lēn) • a simple alkene that is the monomer unit in polyethylene; also called ethene, 112, 125, 128, 129

evaporation (i-va-pə-'rā-shən) • changing from a liquid to a gas; the process by which components of a mixture are separated based on their differences in volatility, 100, 101-102

exchange reaction • see reaction, exchange

F • one-letter abbreviation for phenylalanine, 143

Faraday ('fer-ə-dā), **Michael** (1791-1867 CE) discovered that acids and bases are electrolytes, meaning they form ions when dissolved in water and can conduct electricity, 53

fat • a nutrient that is a major source of energy for living things, is an essential component of cellular membranes, and aids in the absorption of fat-soluble vitamins, 75, 79, 88, 90, 91, 95, 120-121, 136

filter • a device containing holes; used to separate mixtures according to physical size of the components, 98, 100-101

filtration (fil-'trā-shən) • the process of separating components of a mixture based on the differences in their physical size, 100-101

Fischer, Hermann Emil ('fi-shər, 'hər-mən ə-'mēl • [1852-1919 CE] German chemist; proposed ideas about molecular shape for enzymes; developed the Fischer projection, a way of drawing a two-dimensional representation of a three-dimensional molecule, 37

Fischer projection ('fi-shər prə-'jek-shən) • a two-dimensional representation of a three-dimensional molecule where the bonds are drawn as vertical and horizontal lines, 37

flask • a container with a wide bottom and narrow mouth, 15, 16, 17, 102

flask, Erlenmeyer ('ər-lən-mī-er) • a cone shaped flask with a broad, flat bottom and a narrow neck, 16

flask, round-bottomed • a flask with a spherical, not flattened, bottom, 15, 17

flask, volumetric (väl-yü-'me-trik) • a flask with flat bottom and tall neck used to measure a single specific volume, 17

foam • a heterogeneous colloid made of a gas mixed into a liquid or a solid, 91

forceps ('fôr-səps) • a tool used for grasping objects, 14

Franklin, Rosalind • (1920-1958 CE) English chemist; studied DNA using X-ray crystallography; her data was used by Watson and Crick without her knowledge to discover the structure of DNA, 149

free radical ('ra-di-kəl) • a "dangling bond"—an unbonded electron that is very reactive, 128-129

free radical initiator ('frē 'ra-di-kəl i-'ni-shē-ā-tər) • a substance that is added to start a reaction and create free radicals, 128

fructose ('frək-tōs) • a single sugar (a monosaccharide), 80, 117

fume hood • a device used to collect harmful fumes in a chemistry lab, 13

functional group • on a larger molecule, a special site at which chemical reactions occur, 125, 127, 128, 133, 140, 141

Glossary-Index

funnel ('fə-nəl) • a cone-shaped tool with a tube extending down; used to direct the flow of a substance, 14

G • one-letter abbreviation for glycine and for guanine, 143, 150

galactose (gə-'lak-tōs) • a six carbon monosaccharide that is a simple sugar and is part of lactose, 118

gas chromatograph • see chromatograph, gas

genetic (jə-'ne-tik) **code** • the sequence of bases (codons) along the DNA chain that holds the information a cell needs to make proteins and to control its network of machines, 150, 151, 152, 154

glassware • in chemistry, equipment made of glass, 13

gluconeogenesis (glü-kə-nē-ə-'je-nə-səs) • the biochemical pathway in living things that makes carbohydrates when no food is consumed, 79

glucose ('glü-kōs) • a single sugar (a monosaccharide), 80, 81, 82, 117, 118, 119, 120

glutamic acid (glü-'ta-mik 'a-səd) [abbrev., **glu** or **E**] • an acidic amino acid, 141, 142, 143, 154, 155

glutamine ('glü-tə-mēn) • an acidic amino acid that is hydrophilic, 141, 155

gly • three-letter abbreviation for glycine, 143, 155

glyceraldehyde (gli-sə-'ral-də-hīd) • the simplest sugar; a triose with three carbon atoms, 117

glycerol ('gli-sə-rôl) • a small three-carbon carbohydrate, 120

glycine ('glī-sēn) [abbrev., **gly** or **G**] • the simplest amino acid; has an "R" group of one hydrogen atom, 133, 140, 141, 142, 143, 155

glycogen ('glī-kə-jən) • a starch that animals produce in their livers and store in their muscles, 81, 119, 120

gold • an element with 79 protons, 79 electrons, and 118 neutrons, 7, 10, 27, 32, 52, 85-86, 90, 97

graduated ('gra-jə-wā-təd) • marked with lines used for measuring, 17

graduated cylinder ('gra-jə-wā-təd 'si-lən-dər) • a tall glass or plastic cylinder that is used for measuring liquids, 17

graph ('graf) • a diagram of data, 64, 65, 69-70

guanine ('gwä-nēn [abbrev., **G**] • one of the four nucleic acid bases that make up DNA and RNA, 150

Haber (hä-bər), **Fritz** • (1868-1934 CE) a German chemist who tried to create a pH meter, 58

helium ('hē-lē-əm) • an element with 2 protons, 2 electrons, and 2 neutrons, 24, 27

helix ('hē-liks), [plural, **helices** ('hē-lə-sēz)] • a cylindrical or spiral coil shape (like a corkscrew); in a protein secondary structure, the cylindrical coil that forms when the backbone of a polypeptide twists, 119, 144, 146, 151, 152, 156

heptose ('hep-tōs) • a monosaccharide that has seven carbon atoms, 117

heterogeneous (he-tə-rə-'jē-nē-əs) • meaning "other kind;" for mixtures, a mixture that is not the same throughout, 88-91, 97

hexose ('hek-sōs) • a monosaccharide that has six carbon atoms, 117

histidine ('his-tə-dēn) • an amino acid with a five-membered ring, 141, 155

homogeneous (hō-mə-'jē-nē-əs) • meaning "same kind;" for mixtures, a mixture that is the same throughout, 88-91, 97

homopolymer (hō-mə-'päl-ə-mər) • a polymer in which all the repeating monomer units are identical, 125

hydrochloric acid (hī-drə-'klôr-ik 'a-səd) [**HCl**] • a corrosive acid made of hydrogen and chlorine, 48, 55, 56

hydrogen ('hī-drə-jən) • an element that has one proton, one electron, and no neutrons, 9, 27, 28, 38, 40, 46, 47, 48, 53-54, 55, 56, 58, 62, 63, 67, 68, 86, 97, 108, 111, 140, 151

hydrogen ion • see ion, hydrogen

hydrophilic (hī-drə-'fi-lik) • [Gr., *hydro,* water; *philein,* loving] a molecule that will dissolve in water, 94-96, 141

hydrophobic (hī-drə-'fō-bik) • [Gr., *hydro,* water; *phobos,* to fear] a molecule that will not dissolve in water, 94-96, 141

hydroxide (hī-dräk-sīd) • an ion containing one hydrogen atom and one oxygen atom, 47, 53-54, 55, 63

hydroxide ion • see ion, hydroxide

ibn Hayyan, Abu Musa Jabir ('i-bən hā-'yan, 'ä-bü 'mü-sə jä-'bir) • [circa 721-815 CE] Iranian alchemist; discovered hydrochloric acid and aqua regia, 52

imidazole (i-mə-'da-zōl) • a five-membered ring that is the "R" group for histidine, 141

indicator ('in-də-kā-tər) • something that points out, points to, or measures; in chemistry, a substance or device that shows whether a solution is an acid or a base, 57-58, 63-64, 69

infrared spectrometer • see spectrometer, infrared

inorganic chemistry • see chemistry, inorganic

ion ('ī-ən) • an atom that has gained or lost an electron, resulting in a negative or positive charge, 39, 47, 53-56, 58, 63, 68, 87, 89, 91, 93, 143

ion, hydrogen ('ī-ən, 'hī-drə-jən) • H+; a hydrogen atom that has lost its electron and is positively charged, 53-56, 58, 63

ion, hydroxide ('ī-ən, hī-dräk-sīd) • OH-; a molecule made of one oxygen and one hydrogen; has a negative charge, 47, 53-55, 63

ionic bond • see bond, ionic

ketones ('kē-tōnz) • a class of molecules that contain the functional group -C-C=O-C-, 110, 115-116

kinesin (ki-'nē-sən) • a molecular machine that transports cargo along microtubules inside cells, 139, 140

Klemensiewicz, Zygmunt (klə-'men-sə-vich, 'zig-munt) • (1886-1963 CE) Polish physicist who tried to create the first pH meter, 58

lactose ('lak-tōs) • a simple disaccharide sugar molecule found in milk; made of a glucose molecule and a galactose molecule, 118

Lavoisier, Antoine (ləv-'wä-zē-ā, an-'twän) • [1743-1794 CE] a French scientist who did many experiments and discovered that water is made of oxygen and hydrogen, 9, 44

law of conservation of matter • the understanding that atoms are neither created nor destroyed during the process of a chemical reaction, 44

lead (led) • an element with 82 protons, 82 electrons, and 125 neutrons, 7, 10, 27

leucine ('lü-sēn) • a hydrophobic amino acid, 141, 155

Leucippus (lü-'si-pəs) • [c. 480-420 BCE] a Greek philosopher who believed that different atoms have different sizes and weights, 5, 6

lichen ('lī-kən) • an organism consisting of a fungus and algae working in partnership to form the organism; some species are used to make litmus paper, 54

linear polymer • see polymer, linear

lipid ('li-pəd) • any one of a group of nutrients that includes fats, sterols, waxes, fat-soluble vitamins, and other molecules, 120-121

liquid chromatography • see chromatography, liquid

litmus ('lit-məs) **paper** • an acid-base indicator made from paper and dye from lichens, 54-55, 57, 58

lye (lī) • a base used in making soap, 52

lysine ('lī-sēn) • a basic amino acid, 141, 155

macromolecular (ma-krō-mä-'lek-yü-lər) **scale** • particles that are small but much larger than individual molecules; found in homogeneous mixtures, 89

mass spectrometer • see spectrometer, mass

matter • a general term for the stuff that makes up both living and nonliving things, 2, 3-6, 7, 10, 24-29, 43-44, 85, 87, 97

measuring pipet • see pipet, measuring

Mendeleev, Dmitri Ivanovich (men-də-'lā-əf, 'dmē-trē ē-'vän-u-vich) • (1834-1907 CE) Russian chemist who organized the elements into a chart that has been developed into the periodic table of elements, 25-26

messenger RNA ('me-sən-jər 'är 'en 'ā) **(mRNA)** ('em-'är-'en-'ā) • a temporary copy of a gene that tells the ribosome which amino acids to use when constructing a protein, 138, 140, 152, 153, 156

meter, pH • see pH meter

micelle (mī-'sel) • a group of molecules in a surfactant that has both a hydrophobic tail and a hydrophilic head, 95-96

mineral ('min-rəl) • one of a number of the smallest essential nutrients, 76-77, 78

mixture • two or more molecules physically combined but not chemically bonded, 15, 20, 56, 85, 87-104, 112

mobile phase ('mō-bəl 'fāz) • in chromatography, refers to a gas or liquid mixture that has been dissolved in a solvent and flows over a solid, or stationary phase, 103

model ('mä-dəl) • a representation of how something works, 10, 24, 32-37

mole • a group of atoms, molecules, or other things with a quantity equal to 6.022×10^{23}, 68-70

molecular weight (mə-'le-kyə-lər 'wāt) • the combined weight of all the atoms in a molecule; the weight of one molecule, 67, 68

molecule ('mä-li-kyül) • two or more atoms that are chemically bonded together, 32-40, 43-49, 54-56, 58, 62-63, 66-70, 75, 78, 79-82, 86-97, 99, 101-102, 104, 108-122, 124-134, 136-156

monomer ('mä-nə-mər) • a molecule that is an individual unit of a polymer, 119, 120, 124-128, 130-133, 136, 141, 143, 146, 149, 156

monosaccharide (mä-nə-'sa-kə-rīd) • a carbohydrate that is a single, or simple, sugar, 80, 117

mRNA • see messenger RNA

neon ('nē-än) • an element with 10 electrons, 10 protons, and 10 neutrons, 27

neutral ('nü-trəl) • in chemistry, a solution that is neither acidic nor basic, 56, 63

neutral element ('nü-trəl 'e-lə-mənt) • an atom that does not have an electric charge, 24

neutralization reaction • see reaction, neutralization

neutralize ('nü-trə-līz) • to make neutral, 63-64, 68-70

neutron ('nü-trän) • one of the three fundamental particles that make up an atom; found in the nucleus; has an atomic mass of 1 and no charge, 24-25, 27-28, 33, 34, 35, 36, 85

noble gas • one of a group of elements that do not react with other atoms or molecules, 28

N-terminus • the end of a protein that has an amino group, 143

nucleic (nü-'klē-ik) **acid base** • a component of a nucleotide in DNA, 149, 150, 151, 152

nucleic acid polymer (nü-'klē-ik 'a-səd 'pä-lə-mər) • DNA and RNA, 136, 148-153

nucleotide ('nü-klē-ə-tīd) • the monomer unit for DNA, 146, 149, 150, 152

nucleus ('nü-klē-əs) • the center of an atom; made up of protons and neutrons; also called the core, 24, 25, 26, 27, 28, 35, 85

oligosaccharide (ä-li-gō-'sa-kə-rīd) • a simple sugar molecule made of a few saccharides, 118

organic chemist • see chemist, organic

organic chemistry • see chemistry, organic

oxygen ('äk-si-jən) • an element with 8 protons, 8 electrons, and 8 neutrons, 8-9, 26, 27, 38, 40, 44, 46, 47, 48, 67, 76, 82, 85, 86, 87, 90, 92-93, 115

pan scale • see scale, pan

paper chromatography • see chromatography, paper

pelican ('pe-li-kən) • a type of lab glassware used in distillation, 15

pentose ('pen-tōs) • a monosaccharide that has five carbon atoms, 117

peptide ('pep-tīd) **bond** • an amide linkage between amino acids, 141-142, 143

periodic table of elements • a chart used by chemists that categorizes the elements (atoms) and shows their characteristics, 25-29, 67, 108

pH ('pē 'āch) • a measure of the concentration of hydrogen ions in a solution, 56-59

pH ('pē 'āch) **meter** • an instrument that measures pH, 58-59, 63, 64, 69

pH ('pē 'āch) **meter, benchtop** • a stationary pH meter, 59

pH ('pē 'āch) **meter, portable** • a pH meter that can be transported easily in a backpack or pocket, 59

pH ('pē 'āch) **meter, stationary** • an accurate instrument that is used in a lab; also called a benchtop meter, 59

pH ('pē 'āch) **scale** • a measure of how acidic or basic a solution is; has a range of 0 (most acidic) through 14 (most basic), 56, 57

phenolphthalein (fē-nəl-'tha-lēn) • an acid-base indicator, 57

phenylalanine (fe-nəl-'a-lə-nēn) [abbrev., **phe** or **F**] • an amino acid with a six-membered benzene ring as an "R" group, 141, 142, 143, 155

phosphate ('fäs-fāt) **group** • the acid component of a DNA nucleotide, 149, 150, 152

photosynthesis (fō-tō-'sin-thə-səs) • the biochemical process that plants use to convert light energy into food energy, or sugars, 79

physical chemist • see chemist, physical

physical property ('fi-zi-kəl 'prä-pər-tē) • a characteristic that makes atoms and molecules different without changing them chemically, 20, 91, 96-100, 127

pipet (pī-'pet) • a thin tube into which liquid is drawn, 17-18

pipet (pī-'pet), **measuring** • a long, thin tube with volume markings on the side used for measuring liquids, 17

pipet (pī-'pet), **transfer** • a long, thin tube used for moving small amounts of liquids, 18

pipet, volumetric (pī-'pet, väl-yu-'me-trik) • long, thin tube with a bulb in the center used to measure a single specific volume, 17

pipetman (pī-'pet-man) • a handheld piece of equipment that has a little pump inside; used to measure very small amounts of liquids, 18

pleated sheet • in proteins, the sheet-like secondary structure that forms when polypeptide backbones line up next to each other, 144-145

plot • a graphic representation of data, 64-70

point • on a plot, a mark that represents a value or measurement, 65

polar ('pō-lər) • having opposite electrically charged poles, 92-96, 114, 115, 130, 141

polarity (pō-'ler-ə-tē) • the property of having opposite electrically charged poles, 92-93

polyamide (pä-lē-'a-mīd) • a homopolymer, such as nylon, made of identical repeating amide units, 125, 132

polyester (pä-lē-'es-tər) • a common class of condensation polymers formed by hooking monomers together with ester linkages, 133

polyethylene (päl-ē-'eth-ə-lēn) • a homopolymer made from repeating monomer units of a simple alkene called ethylene, 112, 125, 126, 127, 128, 129, 130

polymer ('pä-lə-mər) • a molecule made up of many repeating units, 124-134, 136-157

polymer, amino acid ('pä-lə-mər, ə-'mē-nō 'a-səd) • a protein, 136, 140-148

polymer ('pä-lə-mər), **branched** • a polymer that has side chains of monomer units that form branching structures, 120, 126

polymer, condensation ('pä-lə-mər, kän-den-'sā-shən) • a long-chain polymer formed through a condensation reaction, 132-133

polymer ('pä-lə-mər), **linear** • a polymer that has monomer units connected to each other one after another, end to end, 119, 126, 127

polymer, nucleic acid ('pä-lə-mər, nü-'klē-ik 'a-səd) • DNA and RNA, 136, 148-153

polymerize (pə-'li-mə-rīz) • to build up a polymer through chemical reactions, 127

polysaccharide (pä-lē-'sa-kə-rīd) • a sugar molecule that usually contains ten or more monosaccharides, 80-81, 118, 119, 124

polysaccharide, storage (pä-lē-'sa-kə-rīd, stôr-ij) • a carbohydrate used for storing food energy, 118, 119

polysaccharide, structural (pä-lē-'sa-kə-rīd, 'strək-chə-rəl) • a carbohydrate that is primarily involved in plant cell walls and insect exoskeletons, 118, 119

polystyrene (pä-lē-'stī-rēn) • a homopolymer that is used to make products such as Styrofoam, 125, 127

polyvinyl chloride (pä-lē-'vī-nəl 'klôr-īd) **[PVC]** • a homopolymer; also called PVC, 125

pore • a small opening or hole, such as that in a filter, 100-101

portable pH meter • see pH meter, portable

precipitate (pri-'si-pə-tət) • *noun,* an insoluble substance that results from a chemical reaction, 49, 53, 101

precipitate (pri-'si-pə-tāt) • *verb,* to separate from a solution

Priestley ('prēst-lē), **Joseph** • [1733-1804 CE] an English philosopher and chemist; experimented with oxygen and discovered carbon dioxide gas, 8-9

primary structure • see structure, primary

probability (prä-bə-'bi-lə-tē) • the likelihood that something will take place, 34

probable ('prä-bə-bəl) • likely to happen, 34

product ('prä-dəkt) • an ending material in a chemical reaction, 44, 46, 47, 62

promoter (prə-'mō-tər) • a special control site on DNA where RNA polymerase binds to the DNA, 156

property, chemical • see chemical property

property, physical • see physical property

protein ('prō-tēn) • a polymer made of amino acids; necessary for the overall functioning of biological cells, 75, 81, 88, 90, 104, 108, 115, 133, 136-156

protein domain • see domain, protein

proteosome ('prō-tē-ə-sōm) • a large, complex molecular machine that breaks down proteins, 147, 148

proton ('prō-tän) • one of the three fundamental particles that make up an atom; found in the nucleus; has an atomic mass of 1 and a positive charge, 24-25, 26-28, 33, 35-36, 54, 85

pure substance • an element or a compound, 85-87, 97

PVC • polyvinyl chloride, 125

quantum mechanics ('kwän-təm mi-'ka-niks) • in physics, a theory that uses mathematics to describe the structure and interactions of matter, 34

quaternary structure • see structure, quaternary

R • one-letter abbreviation for arginine, 143

random coil • see coil, random

reactant (rē-'ak-tənt) • a starting material in a chemical reaction, 44

reaction (rē-'ak-shən) • a chemical reaction; occurs whenever bonds between atoms and molecules are created or destroyed, 45

reaction, acid-base (rē-'ak-shən, 'a-səd 'bās) • an exchange reaction between an acid and a base; has a salt as a product, 62-64, 68

reaction, addition • a chemical reaction that links together molecules using double bonds as the functional group, 128-129, 130

reaction, chemical (rē-'ak-shən, 'ke-mi-kəl) • occurs whenever bonds between atoms and molecules are created or destroyed, 45

reaction, combination (rē-'ak-shən, käm-bə-'nā-shən) • a chemical reaction that occurs when 2 or more molecules combine with each other to make a new molecule, 45, 46

reaction, condensation • a chemical reaction in which two monomers combine to form a new molecule, giving off a by-product such as water, 128, 131-134, 142

reaction, decomposition (rē-'ak-shən, dē-käm-pə-'zi-shən) • a chemical reaction that occurs when a molecule decomposes, or breaks apart, into two or more molecules, 45, 46, 48, 62

reaction, displacement (rē-'ak-shən, dis-'plā-smənt) • a chemical reaction that occurs when one atom kicks another atom out of a molecule, 45, 47

reaction, exchange (rē-'ak-shən, iks-'chānj) • a chemical reaction that occurs when one atom in a molecule trades places with another atom, 45, 48, 62

reaction, neutralization (rē-'ak-shən, nü-trə-lə-'zā-shən) • a chemical reaction that occurs when acid molecules react with base molecules; the resulting solution becomes neutral, 63

"R" group • in a chemical formula "R" stands for different types of atoms or functional groups, 140, 141, 159

ribonucleic acid (rī-bō-nü-'klē-ik 'a-səd) • RNA (see RNA), 152

ribose ('rī-bōs) • a monosaccharide that is a pentose with five carbon atoms; a component of a DNA nucleotide, 117, 149, 152

ribosome ('rī-bə-sōm) • the molecular machine that hooks amino acids together; a very large RNA–protein complex that uses mRNA to make a new protein, 138, 140, 142, 152, 153, 156

ring clamp • a ring that can be attached to a pole by means of a clamp; used for holding laboratory equipment such as funnels, 14

RNA ('är 'en 'ā) [ribonucleic acid (rī-bō-nü-'klē-ik 'a-səd)] • a nucleic acid polymer that provides short-term storage of information that is copied from DNA; involved in making proteins, 136, 138, 139, 140, 143, 148, 152-154, 156

RNA polymerase ('är 'en 'ā pə-'lim-ə-rās) • a protein that makes a temporary copy of DNA called messenger RNA (mRNA) which is then used by the ribosome to make new protein, 138, 139, 140, 143, 153, 154, 156

Roberval balance • see balance, Roberval

round-bottomed flask • see flask, round-bottomed

Rutherford, Ernest ('rə-thər-fərd, 'ər-nəst) • British physicist; found that almost all of the mass of an atom is located in the center, and the atom is mainly empty space, 34

safety equipment • objects used in chemistry labs to keep the body safe, 13

saturated ('sach-ə-rāt-əd) **fat** • a fat that has no double bonds in its hydrocarbon "R" groups; animal fat that is solid at room temperature, 120-121

scale • an instrument that measures the weight of an object or substance, 13, 18-20, 66, 68, 69

scale, digital (di-jə-təl) • a scale that has a readout in numerical digits on an illuminated screen, 19

scale, pan • a digital scale with a flat surface where the object being weighed is placed; a top-loading scale 19

scale, top-loading • a digital scale with a flat surface where the object being weighed is placed; a pan scale 19

secondary structure • see structure, secondary

sequence • with reference to proteins, the order of amino acids in a polymer, 143-144, 149, 150, 151, 154, 156

sequence, complementary ('sē-kwəns, käm-plə-'men-tə-rē) • the RNA copy of DNA that is made with complementary base pairs, 154

side chain • a structure that dangles off the main backbone of a polymer, 130, 141

soap • molecules that have both a hydrophobic group (tail) and a hydrophilic group (head), 52, 95-96

sodium ('sō-dē-əm) • an element with 11 protons, 11 electrons, and 12 neutrons, 27, 32, 39, 46, 47, 48, 62, 67, 76, 89, 93, 96, 143

sodium chloride ('sō-dē-əm 'klôr-īd) • NaCl; table salt, 39, 46, 48, 52, 92, 93, 94, 98-99, 101

solid emulsion • see emulsion, solid

solubility (säl-yə-'bi-lə-tē)• the quality of being able to dissolve, 91-93, 97, 99-100, 103

soluble ('säl-yə-bəl) • having the ability to dissolve, 78, 79, 91-93, 98, 99, 104, 120

solute ('säl-yüt) • in a solution, the substance that dissolves, 92

solution (sə-'lü-shən) • a type of homogeneous mixture in which one substance has been dissolved in another, 49, 54-58, 62-64, 69, 88-93, 96, 99

solvent (säl-vənt) • in a solution, the substance that another substance dissolves into, 92, 93, 103

Sörensen, Sören Peter Lauritz ('sô-rən-sən, 'sô-rən 'pē-tər 'lou-rits) • [1868-1939 CE] Danish chemist; introduced the pH scale, 56, 57

spatula ('spa-chə-lə) • a thin, flat tool used for spreading or mixing, 14

spectrometer, infrared (spek-'trä-mə-tər, in-frə-'red) • an instrument used to measure the vibrations of atoms on a molecule, 21

spectrometer (spek-'trä-mə-tər), **mass** • an instrument used to determine the mass of a given sample, 21

mass spectrometer • an instrument used to determine the mass of a given sample, 21

spontaneous (spän-'tā-nē-əs) • occurring without an external influence or force; in a chemical reaction, one that occurs just by mixing chemicals together, 48

starch • a polysaccharide molecule that provides our bodies with energy, 80-81, 82, 88, 115, 119

stationary pH meter • see pH meter, stationary

stationary phase ('stā-shə-ner-ē 'fāz) • in chromatography, a solid over which a gaseous or liquid mixture is passed, causing the components of the mixture to separate, 103-104

steroid ('stir-oid) • a lipid found in both plants and animals; involved in sex hormones, bile acids, and the formation of animal membranes, 121, 136

stir plate • a device used for mixing liquids in a chemistry lab, 13

storage rack • a place to store chemicals in a lab, 13

structure, primary • the sequence of amino acids in a protein chain that determines what kind of machine a protein will become, 143-144

structure, quaternary ('kwä-tər-ner-ē) • a level of protein structure in which there is a combination of two or more protein chains functioning together, 147-148, 156

structure, secondary • the beginning stage of folding in a protein when a helix or pleated sheet is formed, 144-145, 146, 156

structure, tertiary ('tər-shē-er-ē) • the folding of a protein into domains and barrels which occurs after the secondary structure is organized; determines the overall shape of a protein and is critical for protein function, 145-146, 147, 156

sucrose ('sü-krōs) • a disaccharide, 80, 88, 117

surfactant (sər-'fak-tənt) • a molecule that has both a hydrophobic group and a hydrophilic group, 94-96

synthetic (sin-'the-tik) • man-made rather than naturally occurring, 127, 136, 141, 142

synthetic fiber • a man-made polymer that can be drawn out into long, thin fibers, e.g., nylon and Dacron®, 127

T • abbreviation for thymine, 150

tertiary structure • see structure, tertiary

tetrose ('te-trōs) • a monosaccharide that has four carbon atoms, 117

Thales ('thā-lēz) • [625-545 BCE] a Greek philosopher who studied astronomy and mathematics and asked questions about matter; believed that water was the fundamental unit of matter, 3, 6

thermoplastic (thər-mə-'plas-tik) • a polymer that is hard at room temperature but softens when heated, 127, 134

Thomson, Sir Joseph John • [1856-1940 CE] British physicist; studied electricity and how magnetic fields affect the path of light; discovered the electron, 33

thymine ('thī-mēn) [abbrev., T] • one of the four nucleic acid bases that make up DNA, 150

titration (tī-'trā-shən) • the process of using a known solution to find out the concentration of an unknown solution, 63-70

top-loading scale • see scale, top-loading

trace • an extremely small amount, 76

transcription (tran-'skrip-shən) • first step in making a new protein—RNA polymerase copies a piece of DNA, 154

transfer pipet • see pipet, transfer

transfer RNA (tRNA) • a smaller RNA molecule that plays an important role in the synthesis of new proteins, 152, 156

translation (trans-'lā-shən) • the process of copying DNA into a messenger RNA molecule, 156, 157

triglyceride (trī-'gli-sə-rīd) • a derivative of glycerol found in fats, 120

triose ('trī-ōs) • a monosaccharide that has three carbon atoms; the smallest monosaccharides, 117

tRNA • transfer RNA, 152, 156

tube holders • racks and other devices to hold test tubes, 14

tungsten ('təng-stən) • an element that has 74 protons, 74 electrons, and 110 neutrons, 27

unsaturated (ən-'sach-ə-rāt-əd) **fat** • a fat that has double bonds in its hydrocarbon "R" groups; vegetable fat that is liquid at room temperature, 121

uranium (yu-'rā-nē-əm) • an element with 92 protons, 92 electrons, and 146 neutrons, 28

vacuum ('va-kyüm) • empty space containing no matter, 5

valine ('va-lēn) • a hydrophobic amino acid, 141, 155

vitamin ('vī-tə-mən) • an essential molecule found in foods; required for the healthy functioning of living things, 75, 78-79, 120, 136

volatility (vä-lə-'ti-lə-tē) • a substance's ability to evaporate and become a gas, 97, 99, 100, 101-103

volatile (vä-lə-təl) • easily evaporated (changing to the gaseous state), 20, 99, 101-102

volumetric flask • see flask, volumetric

volumetric pipet • see pipet, volumetric

von Hofmann, August Wilhelm • [1818-1892 CE] German chemist; first person to build models of molecules instead of just drawing them, 36

wash bottle • a bottle with a nozzle; used to wash objects in a lab, 14

Watson, James • (born 1928) American molecular biologist; with Francis Crick, determined the structure of DNA, 149

x-axis • the horizontal line on a plot or graph, 65, 70

X-ray crystallography (kris-tə-'lä-grə-fē) • a technique that uses the scattering of radiation to make an image of a large molecule, 149

y-axis • the vertical line on a plot or graph, 65, 70

Pronunciation Key

a	add	j	joy	t	take		
ā	race	k	cool	u	up		
ä	palm	l	love	ü	sue		
â(r)	air	m	move	v	vase		
b	bat	n	nice	w	way		
ch	check	ng	sing	y	yarn		
d	dog	o	odd	z	zebra		
e	end	ō	open	ə	a in above		
ē	tree	ô	jaw		e in sicken		
f	fit	oi	oil		i in possible		
g	go	oo	pool		o in melon		
h	hope	p	pit		u in circus		
i	it	r	run				
ī	ice	s	sea				
		sh	sure				

More REAL SCIENCE-4-KIDS Books
by Rebecca W. Keller, PhD

Building Blocks Series yearlong study program — each Student Textbook has accompanying Laboratory Notebook, Teacher's Manual, Lesson Plan, Study Notebook, Quizzes, and Graphics Package

Exploring the Building Blocks of Science Book K (Activity Book)
Exploring the Building Blocks of Science Book 1
Exploring the Building Blocks of Science Book 2
Exploring the Building Blocks of Science Book 3
Exploring the Building Blocks of Science Book 4
Exploring the Building Blocks of Science Book 5
Exploring the Building Blocks of Science Book 6
Exploring the Building Blocks of Science Book 7
Exploring the Building Blocks of Science Book 8

Focus Series unit study program — each title has a Student Textbook with accompanying Laboratory Notebook, Teacher's Manual, Lesson Plan, Study Notebook, Quizzes, and Graphics Package

Focus On Elementary Chemistry
Focus On Elementary Biology
Focus On Elementary Physics
Focus On Elementary Geology
Focus On Elementary Astronomy

Focus On Middle School Chemistry
Focus On Middle School Biology
Focus On Middle School Physics
Focus On Middle School Geology
Focus On Middle School Astronomy

Focus On High School Chemistry

Super Simple Science Experiments

21 Super Simple Chemistry Experiments
21 Super Simple Biology Experiments
21 Super Simple Physics Experiments
21 Super Simple Geology Experiments
21 Super Simple Astronomy Experiments
101 Super Simple Science Experiments

Note: A few titles may still be in production.

Gravitas Publications Inc.
www.gravitaspublications.com
www.realscience4kids.com